Advanced Textbooks in Control and Signal Processing

T0183758

Series Editors

Professor Michael J. Grimble, Professor of Industrial Systems and Director
Professor Emeritus Michael A. Johnson, Professor of Control Systems and Deputy Director

Industrial Control Centre, Department of Electronic and Electrical Engineering,
University of Strathclyde, Graham Hills Building, 50 George Street, Glasgow G1 1QE, U.K.

Other titles published in this series:

V. Bobál, J. Böhm, J. Fessl and J. Macháček

Digital Self-tuning Controllers

Algorithms, Implementation and Applications

With 187 Figures

 Springer

Vladimír Bobál, Prof. Ing. CSc.
Tomas Bata University in Zlín, Faculty of Technology,
762 72 Zlín, Czech Republic

Josef Böhm, Ing. CSc.
Academy of Sciences of the Czech Republic,
Institute of Information and Automation, 182 08 Praha 8, Czech Republic

Jaromír Fessl, Ing. CSc.
Consultant in Control Engineering, Hornoměcholupská 76, 102 00 Praha 10,
Czech Republic

Jiří Macháček, Doc. Ing. CSc.
University of Pardubice, Faculty of Chemical Technology,
532 10 Pardubice, Czech Republic

British Library Cataloguing in Publication Data
Digital self-tuning controllers : algorithmus,
    implementation and applications. - (Advanced textbooks in control and
    signal processing)
    1. Self-tuning controllers
    I. Bobál, V.
    629.8'36
ISBN-10: 1852339802

Library of Congress Control Number: 2005923880

Advanced Textbooks in Control and Signal Processing series ISSN 1439-2232
ISBN-10 1-85233-980-2
ISBN-13 978-1-85233-980-7
Springer Science+Business Media
springeronline.com

Typesetting: Camera ready by authors
Production and Cover Design: LE-TeX Jelonek, Schmidt & Vöckler GbR, Leipzig, Germany
Printed in Germany
69/3830-543210 Printed on acid-free paper SPIN 11330257

To our wives Jana, Marta, Marie and Marie.

Series Editors' Foreword

The topics of control engineering and signal processing continue to flourish and develop. In common with general scientific investigation, new ideas, concepts and interpretations emerge quite spontaneously and these are then discussed, used, discarded or subsumed into the prevailing subject paradigm. Sometimes these innovative concepts coalesce into a new sub-discipline within the broad subject tapestry of control and signal processing. This preliminary battle between old and new usually takes place at conferences, through the Internet and in the journals of the discipline. After a little more maturity has been acquired by the new concepts then archival publication as a scientific or engineering monograph may occur.

A new concept in control and signal processing is known to have arrived when sufficient material has evolved for the topic to be taught as a specialised tutorial workshop or as a course to undergraduate, graduate or industrial engineers. *Advanced Textbooks in Control and Signal Processing* are designed as a vehicle for the systematic presentation of course material for both popular and innovative topics in the discipline. It is hoped that prospective authors will welcome the opportunity to publish a structured and systematic presentation of some of the newer emerging control and signal processing technologies in the textbook series.

Methods adopted for use in industrial and process control systems are invariably straightforward in structure and easily implemented. The success of industrial PID controllers is often claimed to be due to these factors. The self-tuning controller is a technology, which has all the benefits of structural simplicity and is not very difficult to implement but has not been widely applied in industrial application. One possible reason is the recent extensive development of the robust controller paradigm where the "one size fits all" fixed controller philosophy reigns supreme. Of course, a conservative controller may come with a performance cost degradation so it is always useful to have several tools available for each controller task. And as Professor Bobál and his colleagues Professors Böhm, Fessl and Macháček show in this advanced course textbook the self-tuning controller can be very effective in preserving controller performance in the presence of slowly varying processes, and unknown process disturbances.

Advances on the industrial PID controller will not make the transfer to industrial practice unless there are lucid and direct textbooks available to aid engineers in understanding the potential of these techniques. In this textbook, Professor Bobál and his colleagues have captured their experiences in designing and applying the self-tuning controller method. The book gives a staged presentation that should enable the industrial engineer to develop new industrial applications of this adaptive control technique.

The context of self-tuning controllers is established in the opening three chapters of the book. In these chapters can be found a classification of adaptive control methods establishing the general position of the self-tuning controller method. Chapter 3 serves as an introduction to process model nomenclature and to the techniques of process identification to be used in the text.

Three thorough chapters then follow on different types of control design methods to be used in the self-tuning controller framework. These chapters examine closely the self-tuning PID controller (Chapter 4), the algebraic methods for self-tuning controller design like deadbeat, and pole-placement (Chapter 5) and finally, a self-tuning LQ controller (Chapter 6). Each chapter contains invaluable simulation examples and tips for tuning and implementing the various controller types.

The final two chapters deal with SIMULINK® simulation tools for gaining experience using the self-tuning controllers devised and recount the author team's experiences with some practical process applications. The highlight here is the application of an adaptive LQ controller to a heat-exchanger process.

Since the 1960s, the academic control community has devised many innovative controller methodologies but too few of them have made the transition to regular or widespread industrial practice. This new course textbook on the self-tuning controller method should enable industrial control engineers to gain an insight into the applications potential of this very transparent control technique. The material of the text also gives a good summary of both the theoretical and applications status of the method, which could prove valuable for graduate classes and for re-igniting the method as a research theme.

M.J. Grimble and M.A. Johnson
Industrial Control Centre
Glasgow, Scotland, U.K.
January 2005

Preface

The field of adaptive control has undergone significant development in recent years. The aim of this approach is to solve the problem of controller design, for instance where the characteristics of the process to be controlled are not sufficiently known or change over time. Several approaches to solving this problem have arisen. One showing great potential and success is the so-called self-tuning controller (STC).

The basic philosophy behind STCs is the recursive identification of the best model for the controlled process and the subsequent synthesis of the controller. A number of academics from universities and other institutes have worked intensively on this approach to adaptive control; K. J. Åström (Department of Automatic Control, Lund Institute of Technology), D. W. Clarke (Department of Engineering Science, University of Oxford), P. A. Wellstead (Institute of Science and Technology, University of Manchester), R. Isermann (Department of Control Engineering, Technical University of Darmstadt), I. D. Landau (Institut National Polytechnique de Grenoble), H. Unbehauen (Control Engineering Laboratory, Ruhr University Bochum) and also V. Peterka (Institute of Information Theory and Automation, Academy of Sciences of the Czech Republic, Prague) can be considered as pioneers in this field.

Although during research much effort has been devoted to meeting specific practical requirements it cannot be said that the above approach has been widely applied. On the other hand, many projects have been successfully put into practice. The characteristic common to all these projects was that there was a sufficiently qualified operator available who was both well acquainted with the technology in the field and able to take on board the scientific aspects of the work.

At this current stage of development in adaptive controllers there is a slight growth of interest in both the simpler and more sophisticated types of controller, particularly among universities and companies that deal with control design. It can be seen, however, that the lack of suitable literature in this field imposes a barrier to those who might otherwise be interested. We are referring especially to literature which can be read by the widest possible audience, where the theoretical aspects of the problem are relegated to the background and the main text is devoted to practical issues and helping to solve real problems. In comparison with the most recent publications on this subject, this book leans towards practical aspects, aiming to exploit the wide and unique experience of the authors. An important part of this publication is the detailed documentary and experimental material used to underline the elements in the design approach using characteristics in the field of time or frequency, dealing with typical problems and principles which guide the introduction of individual methods into practice. We should like to note that all the suggested control algorithms have been tested under laboratory conditions in controlling real processes in real time and some have also been used under semi-industrial conditions.

The book is organized in the following way. Chapter 1 gives a brief view of the historical evolution of adaptive control systems. The reader is introduced to problems of adaptive control and is acquainted with a classification of adaptive control systems in Chapter 2. Modelling and process identification for use in self-tuning controllers is the content of Chapter 3. Chapter 4 discusses self-tuning PID (Proportional-Integral-Derivative) controllers. Algebraic methods used for adaptive controller design are described in Chapter 5. Chapter 6 is dedicated to controller synthesis based on the minimization of the linear quadratic (LQ) criterion. Toolboxes have been created for the MATLAB®/SIMULINK® programming system. They serve to demonstrate designed controller properties and help in applications of controllers in user-specific cases. They are described in Chapter 7. Chapter 8 is devoted to practical and application problems. This chapter is based on the rich practical experience of the authors with implementation of self-tuning controllers in real-time conditions.

Although this book is the product of four workplaces (two universities, academia and industry), the authors have tried to take a unified approach. Of course, this has not always been possible. The original work is followed by a list of literature treating the problem under discussion. We assume the reader knows mathematics to technical university level.

This book was created by a team of authors. Chapter 2 was written by V. Bobál, Chapter 3 by V. Bobál together with J. Böhm. V. Bobál and J. Fessl created Chapter 4 as follows: Sections 4.1 and 4.2 they wrote together, Sections 4.3, 4.4, and 4.5 are by J. Fessl, and Sections 4.6, 4.7, 4.8, and 4.9 are by V. Bobál. J. Macháček and V. Bobál wrote Chapter 5. Chapter 6 was written by J. Böhm and Chapter 7 by V. Bobál and J. Böhm. Finally Chapter 8 is a corporate work by all authors.

This book appears with the support of the Grant Agency of the Czech Republic, which provided the funding for projects numbered 102/99/1292 and 102/02/0204 and by the Ministry of Education of the Czech Republic under grant No. MSM 281100001.

We would like to thank our colleagues and students in the Institute of Process Control and Applied Informatics, Faculty of Technology at the Tomas

Bata University in Zlín for their assistance in the preparation of toolboxes and the camera-ready manuscript, namely Dr Petr Chalupa, Dr František Gazdoš, Dr Marek Kubalčík, Alena Košťálová and Jakub Novák.

We would finally like to thank the Series Editors Professor M. J. Grimble and Professor M. A. Johnson for their support during the publication of this book.

Zlín, Praha, Pardubice
December 2004

Vladimír Bobál
Josef Böhm
Jaromír Fessl
Jiří Macháček

Contents

1

Introduction

For over forty years the field of adaptive control and adaptive systems has been an attractive and developing area for theorists, researchers, and engineers. Over 6 000 publications have appeared during the history of adaptive systems and this number is certainly not definitive. However, the number of industry applications is still low because many manufacturers distrust nontraditional and sometimes rather complicated methods of control. The classic methods of control and regulation have often been preferred. These have been worked out in detail, tested, and in many cases reached the desired reliability and quality. On the other hand, it is necessary to realize that the vast majority of processes to be controlled are, in fact, neither linear nor stationary systems and change their characteristics over time or when the set point changes. Such changes affect different processes in various ways and are not always significant. In other systems and processes, however, the changes may be significant enough to make the use of controllers with fixed parameters, particularly of a PID type, unacceptable or eventually impossible.

The first attempts to develop a new and higher quality type of controller, capable of adapting and modifying its behaviour as conditions change (due to stochastic disturbances), were made in the 1950s. This was in the construction of autopilot systems, in aeronautics, in the air force, and in the military. The concept of adaptation that in living organisms is characterized as a process of adaptation and learning was thus transferred to control, technical, and cybernetic systems.

From the beginning the development of adaptive systems was extremely heterogeneous and fruitful. The results were dependent on the level of the theory used and the technical and computing equipment available. During the first attempts simple analogue techniques (known as MIT algorithms) were applied. Later algorithms became more complicated and the theory, more demanding. Many approaches were not suitable for real-time applications because of the performance of the available computers or were simply too sophisticated for analogue computers. That period brought results of a theoretical research value only. Reference [1] from 1961 shows the wealth of methods and approaches dating from this pioneering age.

During the 1960s two main areas emerged from this diversity of approaches to dominate the field of adaptive systems for many years. The first were Model Reference Adaptive Systems (MRAS) in which the parameters of the controller modify themselves so that the feedback system has the required behaviour. The second were self-tuning controllers (STC), which due to use of the matrix inversion lemma were able to use measured data to identify the model (the controlled process) on-line. The linear feedback controller parameters adapt according to the values of the identified parameters of the process model. Naturally both directions had their own supporters and developments were outlined in [2]. The next decade was characterized by growing attempts to use adaptive systems in real-world applications, increased use of modern computers, and by applying the latest information and methods available in the theory of control. Examples include:

- the use of algebraic approach in control design,
- the parameterization of controllers,
- the use of rational fraction functions,
- the digitalization of signals and models.

A survey of developments during this period is given in [3]. The 1980s saw further breakthroughs. As microprocessor technology became ever faster, cheaper and more compact, and analogue equipment began to fall into disuse, the level of digitalization increased and became attractive for real-time use. Adaptation was also relevant to other areas such as filtration, signal prediction, image recognition, and others (see [4]). Methods known as auto-tuning started to appear, in which adaptation only occurs in the first stage of control (in order to identify the right controller type, which then remains fixed). Conferences dealing specifically with adaptive systems were held and the number of monographs and special publications increased. The developments that took place during this period is documented in [5]. In the early 1990s new discoveries were applied to adaptive methods, such as artificial intelligence, neuron networks and fuzzy techniques.

However, in the second half of the 1990s adaptive systems still showed great unused potential in mass applications even though many well-known companies deployed adaptive principles for auto-tuning and occasionally even for on-line control. There were still opportunities for improvements, for streamlining in the areas of theory and application, and for increasing reliability and robustness. It has been accepted that there are processes that can only be controlled with automatic adaptive controllers but that are still controlled manually. In many real-world processes this high quality control must be ensured and as the process alters, this leads back to adaptation. It is reasonable to assume that even where a nonadaptive controller is sufficient, an adaptive controller can achieve improvements in the quality of control. An example of this is given in [6] where the use of an adaptive controller decreased fuel consumption significantly.

This publication is devoted to adaptive controllers of the STC type. On-line process identification is followed by a controller design. Various design techniques can be used. PID, algebraic methods and the LQ approach to controller design are considered. Specific examples of solutions are provided including their simulation environment. This discussion is suitable for students of engineering and those working on theses in the field of technical cybernetics and automation at universities of technology.

2

Adaptive Control Systems

The majority of processes met in industrial practice have stochastic character. Traditional controllers with fixed parameters are often unsuited to such processes because their parameters change. Parameter changes are caused by changes in the manufacturing process, in the nature of the input materials, fuel, machinery use (wear) *etc.* Fixed controllers cannot deal with this.

One possible alternative for improving the quality of control for such processes is the use of adaptive control systems, which has been made possible by the development of modern digital automation based on microprocessor technology. Naturally this must be taken together with the development and improvement of adaptive control algorithms, and the exploration of their potential, advantages and limitations.

This chapter is divided into two main sections followed by a summary in Section 2.3. Formulation of the adaptive control problem is introduced in Section 2.1, and classification of adaptive control systems from the point of view of basic approach in Section 2.2.

2.1 Formulation of Adaptive Control Problem

Originally, adaptation was displayed only by plants and animals, where it is seen in its most varied forms. It is a characteristic of living organisms that they adapt their behaviour to their environment even where this is harsh.

Each adaptation involves a certain loss for the organism, whether it is material, energy or information. After repeated adaptations to the same changes, plants and animals manage to keep such losses to a minimum. Repeated adaptation is, in fact, an accumulation of experiences that the organism can evaluate to minimize the losses involved in adaptation. We call this learning.

Alongside such systems found in nature there are also technical systems capable of adaptation. These vary greatly in nature, and a wide range of mathematical tools are used to describe them. It is therefore impossible to find a single mathematical process to define all adaptive systems. For the purposes of our definition of adaptive systems we will limit ourselves to cybernetic systems which meet the following assumptions:

- their state or structure may change;
- we may influence the state or output of the system.

One possible generalized definition of an adaptive system is as follows:

The adaptive system has three inputs and one output (Figure 2.1). The environment acting on the adaptive system is composed of two elements: the reference variable w and disturbance v. The reference variable is created by the user but, as a rule, the disturbance cannot be measured. The system receives information on the required behaviour Ω, the system output is the behaviour of the system (decided rule)

$$y = f(w, v, \Theta) \tag{2.1}$$

which assigns the single output y to each behaviour occurring in environments w and v. A change in behaviour, *i.e.* a change in this functionality, is effected by changing parameters Θ. For each combination (w, v, Θ) we select in place of Θ parameter Θ^* so as to minimize loss function g (for unit time or for a given time period)

$$g(\Omega, w, v, \Theta^*) = \min g(\Omega, w, v, \Theta) \tag{2.2}$$

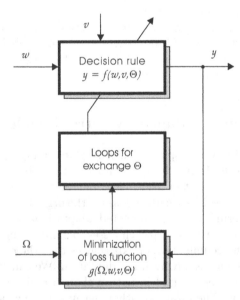

Figure 2.1. Inner structure of an adaptive system

In this case adaptation is the process used to search for $\boldsymbol{\Theta}^*$ and continues until this parameter is found. A characteristic property of an adaptive system is the fact that the process of adaptation always occurs when there is a change in the environment w or v or a change in the required behaviour Ω. If a change occurs after each time interval T_0 adaptation will take place repeatedly at the start of each interval. If the adaptation then lasts for time τ (after which loss g decreases) then the mean loss will be lower with a smaller ratio τ/T_0. The inverse value of the mean loss is known as the adaptation effect.

We mention here the so-called learning system. The learning system can be seen as a system that remembers the optimal value of parameter $\boldsymbol{\Theta}^*$ on finishing the adaptation for the given m triplet (w_m, v_m, Ω_m) of sequence $\{(w_m, v_m, \Omega_m)\}$, for $k = 1, 2, \ldots, m, \ldots, \infty$, and uses it to create in its memory the following function

$$\boldsymbol{\Theta}^* = f(w, v, \Omega) \tag{2.3}$$

On completing the learning process the decided rule for every behaviour in environments w and v can be chosen directly by selecting the appropriate value for parameter $\boldsymbol{\Theta}^*$ from memory without adaptation.

We can conclude, therefore, that an adaptive system constantly repeats the adaptation process, even when the environment behaviour remains unchanged, and needs constant information on the required behaviour. A learning system evaluates repeated adaptations so as to remember any state previously encountered during adaptation and when this reoccurs in the environment, does not use Equation (2.2) to find the optimum but uses information already in its memory.

Adaptive and learning systems can be used to solve the following tasks:

- recursive identification – *i.e.* the creation of a mathematical description of the controlled process using self-adjusting models;
- the control of systems about which we know too little before starting up to predefine the structure and parameters of the control algorithm, and also systems whose transfer characteristics change during control;
- recognition of subjects or situations (scenes) and their classification. Adaptive and learning systems are then components of so-called classifiers;
- manipulation of subjects – *i.e.* change of their spatial position. Adaptive and learning systems are then components of robots.

Further we will focus only on problems of adaptive control. Figure 2.2 shows a general block diagram of an adaptive system. According to this diagram we can formulate the following definition:

An adaptive system measures particular features of the adjustable system behaviour using its inputs, states and outputs. By virtue of comparison of these measured features and sets of required features it modifies parameters and the structure of an adjustable loop or generates an auxiliary input so that the measured features track as closely as possible the required features.

This definition is fairly general and allows inclusion of most of the adaptive problems of technical cybernetics. Features of the behaviour can take different forms in these problems. If the adaptive system is used for control, the behaviour feature could be, for example,

- pole and zeros assignment of a closed loop system;
- the required overshoot of the step response of a closed loop system to reference and input disturbances;
- the settling time;
- the minimum value of various integral or summing criteria;
- the amplitude and natural frequency of oscillations in nonlinear loops;
- the frequency spectrum of a closed loop control system;
- the required value of gain and phase margins *etc.*

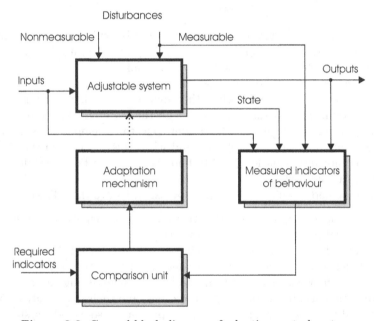

Figure 2.2. General block diagram of adaptive control system

For the purposes of automatic control we can simplify the definition of an adaptive system still further:

Adaptive control systems adapt the parameters or structure of one part of the system (the controller) to changes in the parameters or structure in another part of the system (the controlled system) in such a way that the entire system maintains optimal behaviour according to the given criteria, independent of any changes that might have occurred.

Adaptation to changes in the parameters or structure of the system can basically be performed in three ways:

- by making a suitable alteration to the adjustable parameters of the controller;
- by altering the structure of the controller;
- by generating a suitable auxiliary input signal (an adaptation by the input signal).

2.2 Classification of Adaptive Control Systems

The difference between classic feedback controllers and adaptive controllers is that the classic controller uses the principle of feedback to compensate for unknown disturbances and states in the process. Feedback is fixed and amplifies or otherwise modifies the error $e = w - y$ (w is the reference value of process output y), which in turn determines the value of the input signal u (controller output) for the system. The way in which the error is processed is the same in all situations. The basis of the adaptive system is that it alters the way in which the error is processed, *i.e.* adapts the control law to unknown conditions and extends the area of real situations in which high quality control can be achieved. Adaptation can be understood as feedback at a higher level where the controller parameters change according to the quality of the control process.

In recent years the theory of adaptive control has made significant developments. Obviously, as in any other new scientific discipline, the theory of adaptive control has no unified approach to classifying the systems operating on this principle. Here it suffices to use the classification set out in Figure 2.3; learning systems are not included. For detailed adaptive control systems classification according to different approaches see [7].

Adaptive systems based on the *heuristic approach, self-tuning controllers* (STC) and *model adaptive reference systems* (MRAS) are currently the three basic approaches to the problem of adaptive control. Adaptive systems which have a variable structure will purposely alter their structure following the set procedure. Since such a system alters its structure on the basis of experience gained previously in its working life, it may be regarded as a *self-organizing system*.

2.2.1 Adaptive Controllers Based on a Heuristic Approach

Methods using this approach provide adaptability directly either by evaluating the process output (or its error) or selected quality criteria for the control process. In these cases the algorithm for a PID digital controller is often used and we usually select the level of oscillation in the process output, or its error, as the criterion. These methods do not require identification of the controlled system. In some cases it is not even necessary to monitor the output error or introduce special test signals. A block illustration of these methods is given in Figure 2.4. Process output y, or error e, are evaluated according to the

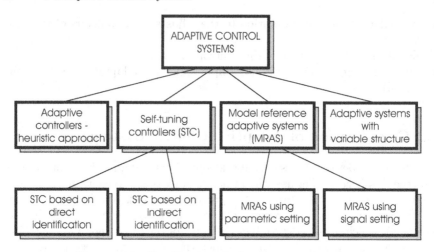

Figure 2.3. Classification of adaptive control systems

supplied criterion and subsequently preset the parameters of a PID controller.

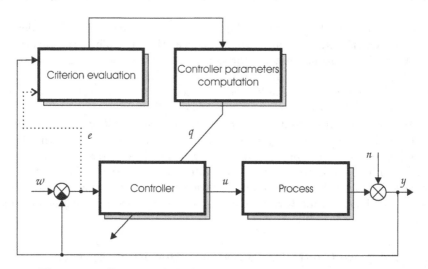

Figure 2.4. Diagram of the heuristic approach to adaptive control

When synthesizing this kind of controller we try to optimize the criterion that quantifies the quality of the control process. Although this approach satisfies practical applications while also being robust, it comes up against a number of calculation problems and has only been successfully applied in the simplest cases.

One of these successful applications is the approach designed by Maršík [8]; the method involves setting the gain of the PID controller and selecting the oscillation range as the directly measurable criterion. We know that the control process oscillates more as it approaches the limits of stability, while too damped a process ceases to oscillate at all. There are several modifications to this type of controller, some of which are so simple that they can even be realized by a few dozen fixed point operations.

Åström and Hägglund added the term self-tuning alongside the expression auto-tuning [9]. Although auto-tuning cannot be regarded as the same as self-tuning we will describe some of the principles of these controllers in this section. One example of this type of controller which has been fairly widely applied in practice is the auto-tuning controller designed by Åström and Hägglund [10, 11], where alongside the PID controller a relay type of nonlinearity is inserted in parallel into the feedback. During the adjustment phase the relay is introduced into the feedback causing the control loop to oscillate at a critical frequency. Since controller output u acquires only two values $\pm R$ and is therefore a rectangular process, the process output y has an approximately sine wave pattern, the shape of which depends on how the system filters harmonics out of the controller output. A simple Fourier series expansion of the relay output shows that the first harmonic component has amplitude $4R/\pi$. The ultimate (critical) gain K_{Pu} is then given as the ratio of the amplitude of the first harmonic component and the error amplitude $e_{\max}$

$$K_{Pu} = \frac{4R}{\pi e_{\max}} \tag{2.4}$$

The ultimate (critical) period T_u is measured from the cycling.

The controller can also be automatically adjusted by evaluating transient processes. Kraus and Myron [12] describe the so-called "EXACT Controller" (Expert Adaptive Controller Tuning) from the company Foxboro. This auto-tuning controller uses the pattern recognition approach, *i.e.* knowledge of the error process during transition. To make adjustments the controller uses three peaks from the error process to calculate overshoot and undershoot, which in turn is used together with the oscillation period to set the parameters for the PID controller. Some authors (such as Nishikawa *et al.* [13]) suggested adjusting PID controllers by measuring system response of the reference signal or process output in an open or closed loop. The parameters of the PID controller are optimized by calculating the integral linear or quadratic criterion for control process quality. In recent years much has been published on auto-tuning controllers, especially of the PID type (see [14, 15] and others).

2.2.2 Model Reference Adaptive Systems

The problem of model reference adaptive systems design is theoretically well elaborated and widely discussed in the scientific literature [16, 17]. The basic block diagram of the model reference adaptive system is shown in Figure 2.5.

The reference model gives requested response y_m or requested state vector x_m to reference input signal u_r.

This approach is based on observation of the difference between the output of the adjustable system y_s and the output of the reference model y_m.

The aim of the adaptation is convergence of the static and dynamic characteristics of the adjustable system, *i.e.* the closed loop, to the characteristics of the reference model. This, in fact, is an adaptive system with forced behaviour where the comparison between this forced behaviour and the behaviour (response) of the adjustable system (control loop) y_s, provides the error ε. The task of the appropriate control mechanism is to reduce error ε or errors in the state vector x between the reference model and the adjustable system to a minimum for the given criteria. This is done either by adjusting the parameters of the adjustable system or by generating a suitable input signal, as can be seen in Figure 2.5.

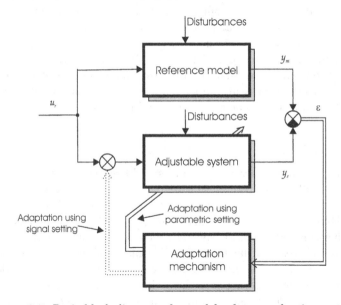

Figure 2.5. Basic block diagram of a model reference adaptive system

The dual character of this adaptive system is important since it can be used both for control and to identify the parameters of the model process or to estimate the state of the system. These systems are, to a certain extent, limited by the fact that they are only suited to deterministic control, however prospects for their wider use are good.

2.2.3 Self-tuning Controllers

For the two approaches previously outlined the design of an adaptive controller did not require detailed knowledge of the dynamic behaviour of the controlled system. Another approach to adaptive control is based on the recursive estimation of the characteristics of the system and disturbances and updating the estimates, so monitoring possible changes. Using this knowledge, appropriate methods can be employed to design the optimal controller. This kind of controller, which identifies unknown processes and then synthesizes control (adaptive control with recursive identification) is referred to in the literature as a self-tuning controller – STC. The most useful results in practical terms have been achieved mainly in one-dimensional systems for which a number of numerically stable algorithms of varying complexity have been designed. These algorithms can then be applied via a control computer equipped with a unit to interface with the technological environment. Expanding this to multivariable systems in many cases does not cause fundamental problems.

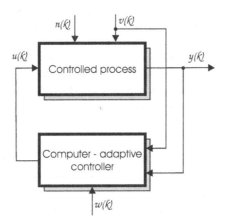

Figure 2.6. Basic block diagram of digital adaptive control loop

We assume a controlled technological process with a single process input $u(k)$ and a single process output $y(k)$. In addition, measurable disturbance $v(k)$ and nonmeasurable disturbance $n(k)$ – random noise – may affect the controlled process. A computer working as the digital adaptive controller is connected in feedback to the controlled process and, among other things, processes the required value of the process output. The block diagram of this basic feedback loop is given in Figure 2.6.

An adaptive digital controller works with a fixed sampling period T_0. A controller with this period generates a sequence of numerical values for the controller output $\{u(k); k = 1, 2, \ldots\}$ (assuming $T_0 = 1$). The discrete controller output $u(k)$ operates via an digital–to–analogue (D/A) converter and actuator in the closed loop. The controller output value is constant during the sampling interval. The actuator, including D/A converter, is included in the dynamics of the controlled process. The output of the controlled process is a physical (usually continuous-time) variable, which is also sampled over period T_0. Therefore, as far as the controller is concerned, the process output is a sequence of numerical values $\{y(k); k = 1, 2, \ldots\}$ and this is the only information the controller has regarding the continuous-time output. It can sometimes be useful to filter the continuous-time system output before sampling. The sensor, A/D converter and any filter being used are also regarded as part of the controlled process.

The basic feedback loop can be extended by a forward loop from the externally measured disturbance $v(k)$, if such measurements are available. Its behaviour is also sampled over period T_0 and transferred to the controller as a sequence of numerical values $\{v(k); k = 1, 2, \ldots\}$, and at the same time the reference variable value is digitally expressed as a sequence of numerical values $\{w(k); k = 1, 2, \ldots\}$. The existence of random nonmeasurable disturbance $v(k)$ and any change in the reference variable value $w(k)$ is the reason for introducing automatic control. The aim of control is to compensate these disturbances as well as to track the reference variable values. Further, we assume that the parameters of the controlled process are either constant but unknown or variable, in which case changes in these parameters are significantly slower than the speed of the adaptation process. Depending on the nature of the controlled process we can see how the following aims can be achieved using adaptive control with recursive identification:

- automatic tuning of the digital controller;
- improved control where nonstationary disturbances are present;
- the detection of changes in the parameters of the controlled system arising from various technological causes, for example changing the operating mode of the equipment;
- improvement in the control procedure of a given process by making a suitable change to the parameters of the digital controller.

Algorithmic Structure of Self-tuning Controllers

It is clear that to reach these goals the identification of the static and dynamic characteristics of a given process plays an important role together with the optimal control strategy itself. From parameter estimation theory we know that the determination of parameters is always burdened by a degree of uncertainty – error. This uncertainty not only depends on the number of identified steps (*i.e.* on the amount of sampled data) and on the choice of structure for the mathematical model of the controlled process, but is also dependent on the behaviour of the controller output, the sampling period and the choice of filter for the controller and process outputs. This means that every realized change in a controller output except the required control effect, also excites

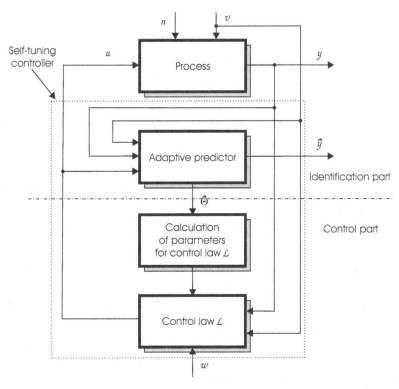

Figure 2.7. Internal algorithmic structure of a self-tuning controller

the controlled system and thus creates the conditions for its identification; in other words, for the best identification of the controlled process, it is necessary to impose certain conditions on the course of controller outputs.

The general task of optimal adaptive control with recursive identification is, therefore, extremely complicated because we have to seek within it a sequence of such controller outputs which can ensure that the mean process output value is as close as possible to the target value and at the same time enable the most accurate identification of the given process. Feldbaum [18] has presented a design for optimal control which, to a certain extent, fits the given assumptions. Because this design for optimal control has two effects it is called dual optimal control. Unfortunately, due to the complexity of the calculations it involves, dual optimal control is too demanding to be of use in most situations. Although exceptional efforts have been devoted to dual optimal control, not even the use of various simplified approximations has managed to reduce it to a stage where it can be applied practically.

It has, therefore, been necessary to simplify the solution to this problem using experimental experience and intuition. This solution is called forced separation of identification and control – the Certainty Equivalence Principle. The principle of this simplification is outlined in the following procedure:

1. The vector of process model parameters Θ are regarded as being known for each control step and this equals its point estimate, which is available at any given moment, *i.e.* $\Theta = \hat{\Theta}(k-1)$.
2. The design of the control strategy to affect the desired control quality criteria is based on this assumption and the required controller output $u(k)$ is calculated.
3. Having acquired a new sample of process output $y(k)$ (or external measured disturbance $v(k)$) and known controller output $u(k)$ a further step in identification is performed using a recursive identification algorithm. This means that the new information on the process deduced from the three data items $\{u(k), y(k), v(k)\}$ is used to update estimate $\hat{\Theta}(k-1)$ and the entire procedure is repeated to make a new estimate $\hat{\Theta}(k)$.

From experience gained during experimentation we can see that the majority of practical tasks of adaptive control with recursive identification are suited to the given simplified approach.

The approach described above implies the inner algorithmic structure of the self-tuning controller schematically shown in Figure 2.7. The forced separation of identification and control splits the inner controller structure into parts for identification and control, which are only connected through the transfer of point parameter estimates $\hat{\Theta}(k)$. Recursive estimation of the process model parameters is carried out in the identification part and used to predict value $\hat{y}(k)$ of process output $y(k)$. The control part contains a block to calculate the control parameters (control law L) using the process model parameter estimates $\hat{\Theta}(k)$. The control parameters then serve to calculate the value of controller output $u(k)$ for each sampling period.

As can be seen from the structure above, reliable and quickly convergent identification is absolutely vital if the controller is to function well. Even though certain specific conditions are applied to the synthesis of adaptive control we can state that, where identification works well, synthesis can be carried out using known algorithms such as those for pole assignment design, dead-beat control, minimum variance control, generalized minimum variance control, linear quadratic control, and digital synthesis methods for PID controllers. The STC algorithms mentioned in this monograph differ only in the control path; for identification we will be using the recursive least squares method.

In some STCs the identification process does not serve to determine estimates of the process model parameters $\hat{\Theta}(k)$, rather, appropriate reparametrization of the control loop can be used recursively to estimate the controller parameters directly. This means it is necessary to find the relationship between the process input and output and define it directly from the controller parameters without recalculating them using the estimates of the process model parameters. These controllers are referred to as being *implicit*,

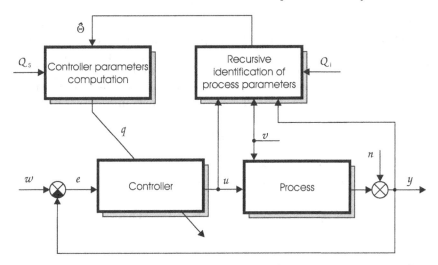

Figure 2.8. Block diagram of an explicit STC (with direct identification)

whereas controllers using a synthesis from estimates of the process model parameters are called *explicit*. If we illustrate an explicit STC using a diagram like the one in Figure 2.8, which is analogous to Figure 2.7, where Q_i is the identification criterion, Q_s the controller synthesis criterion and q are the controller parameters, we may draw a diagram of an implicit STC as in Figure 2.9.

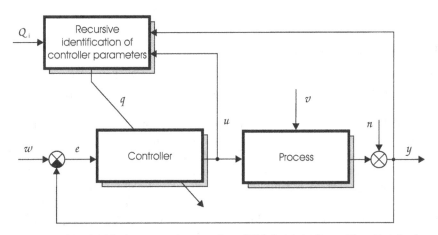

Figure 2.9. Block diagram of an implicit STC (with indirect identification)

The STC principle can also be used for one-shot controller tuning (autotuning). If the algorithm illustrated in Figure 2.7 is used for controller autotuning then the blocks representing recursive identification and controller parameter calculation are only connected at the moment when the controller is being set up, *i.e.* during the adjustment phase. Once the controller has been adjusted they are disconnected. The system is then controlled by fixed parameters. Clearly, this method of control is useful for deterministic processes where identification is switched off once the controller has been adjusted. The underlying principle of this type of controller is shown in Figure 2.10.

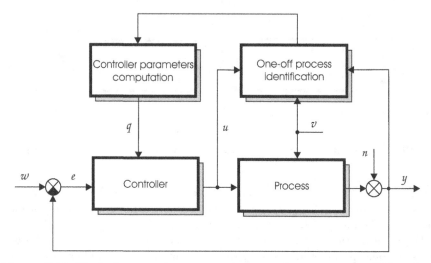

Figure 2.10. Block diagram of an auto-tuning controller using a one-off identification process

Development of Self-tuning Controllers

Here we give a brief history of the development of explicit STCs. The approach used in STCs was first mentioned in the work of Kalman [19] in 1958. He designed a single-purpose computer to identify the parameters of a linear model process and subsequently calculate the control law using minimum quadratic criteria. This problem was revived in the early 1970s by the work of Peterka [20] and Åström and Wittenmark [21] and others. The approach has been developed significantly since then. The first STCs were designed so as to minimize system output variance where some disadvantages were removed by the general minimization of output dispersion method developed by Clarke and Gawthrop [22, 23]. These are known as single-step methods because only one sample of the process output is considered in the quadratic criterion. One great disadvantage is that they are unable to control the so-called nonminimum phase systems, which are processes where the polynomial $B(z)$ (the polynomial which is included in the controlled system transfer function numerator) has its poles outside the unit circle of the complex z-plane, *i.e.* in an unstable area.

This problem can be solved by using multi-step criteria (to limit an infinite number of steps) which is a solution to the general quadratic problem. Peterka analyzed the probability rate of the Bayessian approach to adaptive control based on linear quadratic synthesis [24]. In general this control synthesis involves rather complex iteration calculations [25]. It has been shown that analytic methods may be used to find relatively simple explicit relationships to determine the optimal controller for one-dimensional models that are no higher than second order (see Böhm *et al.,* [26]).

In the late 1970s, early 1980s the first work was done on STCs based on pole assignment [27, 28]. During the 1980s much attention was also paid to single- and multi-step prediction adaptive methods [29, 30]. Hybrid STCs using a δ operator have similarly been analyzed [31]. Alongside these developments there has been exploration into synthesis methods for digital PID controllers which might be able to use parameter estimates gained through recursive identification to calculate controller intervention. These parameters are then used to calculate the PID controller elements, that is gain K_P, and integral and derivative time constants T_I and T_D [32, 33].

Problems for Further Research of Self-tuning Control Design

The principles of adaptive controllers bring also some drawbacks, mainly in the area of reliability, a property that is very important for any application. The problem is caused mainly by the identification part. Conditions for unbiased estimation cannot always be satisfied. Most of the present adaptive controllers consider only model parameter adaptation. Thus adaptive controllers are suitable only for slowly varying processes. In cases when parameter changes are abrupt, *e.g.* in the case of nonlinearities, fault states or rapid changes of process working conditions, parameter adaptivity cannot react properly.

To improve the identification part, developments in the following areas may be promising:

- use of different, usually more complex, identification methods which have less strict conditions for correct estimation;
- use a set of fixed models describing the given plant instead of on-line identification;
- use of a supervisor, monitoring controller behaviour and correcting its behaviour.

The idea of a supervisor deserves further attention. A supervisor can be implemented simply and some forms of it are now frequently used. Any practical application is completed by a process-specific supervisor that switches the adaptive controller over to an available standard controller in the case of unexpected controller behaviour.

A data-dependent forgetting factor in identification is now standard [34]. Data-dependent controller synthesis is used in Chapter 6 and Chapter 8.

More complex supervisors can be developed based on artificial inteligence approaches. Methods such as fuzzy logic, evolutionary algorithms [35, 36], or their combination show themselves to be promising methods from the artificial intelligence field [37, 38]. An intelligent supervisor based on these methods can be used. Its task generally is parameter adjusting and/or choice of an appropriate controller. If a fuzzy supervisor is used, then it can be regarded as a set of static nonlinear I/O functions. Its task is controller parameter adjustment if needed, and/or the choice of an appropriate controller from classes of controllers. Adjustment of the fuzzy controller must be done for error-free function. Today, there are three common methods for this. In the first, an expert adjusts all fuzzy logic properties according to his experience; in the second, some sophisticated method is used [39]; and finally in the third, a fuzzy supervisor is adjusted on-line by means of modern evolutionary algorithms [35]. If the evolutionary algorithm is used like an evolutionary supervisor, i.e. independently of other methods, then its entire task can be regarded as a nonlinear constrained multi-objective optimization problem. An advantage of such a supervisor is its simplicity, universality, and of course, its independence from a human operator. Nowadays, similar methods are applied in intelligent sensors for the class of so-called fault-detection tasks.

2.3 Summary of chapter

This chapter gives a formulation of the adaptive control problem, and defines those problems that can be solved using adaptive systems. A simplified definition of the adaptive system suitable for the design of adaptive control loops is also given. Adaptive control systems can be classified as adaptive controllers based on heuristic approach, model reference adaptive systems, and self-tuning controllers.

The principles of the self-tuning controller are described including its algorithmic structure. The differences between implicit and explicit versions of the self-tuning controller are also explained. The chapter concludes with a summary of the historical development of self-tuning controllers and describes further developments in this area of adaptive control.

Problems

2.1. Clarify the general term "adaptivity" and utilization of the principle for process control.

2.2. Describe a class of technological processes suitable for the implementation of adaptive controllers.

2.3. Design a block diagram of an MRAS for adaptive system identification.

2.4. What are the differences between explicit and implicit versions of STC?

3

Process Modelling and Identification for Use in Self-tuning Controllers

Controller designs rely on knowledge of the controlled plant. Since complete knowledge is illusory we refer to knowledge of the plant model or controlled process model. Introducing the term model explicitly expresses the potential difference between the reality of the controlled process and the abstract mathematical model (in specific fields, other types of models may also be used).

The aim of the model is to give a faithful representation of the process behaviour. However, the term "faithful" can be interpreted in many different ways according to the purpose for which the model is being used. Here the model is used to design a controller, so by the word "faithful" we mean that the controller designed for the model under consideration will also operate with a real plant. Because we will be dealing with several design methods it is also necessary to take the specific faithfulness of each method into account.

Traditionally it is stated that a model can be derived from mathematical analysis of the physical and chemical processes in the plant or from an analysis of measured data. Adaptive control models are mainly represented by the second approach, although, as we will point out in the section dealing with identification, the first approach cannot be completely omitted.

As was shown in Section 2.2.3, an algorithm for recursive identification of the parameters of the process model is an indispensable part of every self-tuning controller. Consequently, it is obvious that a model obtained by one-off processing of system data is not sufficient. Therefore, when selecting a model for adaptive control purposes, it is necessary to consider the following assumptions:

1. The controlled object must be identifiable, which means that it is possible to model it by analysis of measurable values of input and output variables.
2. The model must be selected from the class of so-called parametric models. These are models that can be described as a function of independent variables (in the discrete-time version using samples of the last values of input and output variables) and from a finite number of parameters. Determination of these parameters is the subject of identification.

3. A criterion for comparing the differences between various types of models from the given class must be chosen. It has to be defined by a relation allowing one to compare measured data with model data. For the criterion, it is necessary to fulfil the following conditions:
   - it must be a suitable measure for detecting the differences between model and object behaviour;
   - it must be sensitive to random effects and errors;
   - it must be easily computable.

The aim of this chapter is to introduce the reader to the problems of recursive identification of processes with regards to practical aspects of self-tuning controllers design. Selection of the stochastic process model in ARX and ARMAX structures is discussed briefly. For practical purposes the ARX model is chosen, and to obtain parameter estimates of the one-off model the least squares method is introduced. A recursive algorithm for this method is also derived. In order to ensure numeric stability of the computation, the square root filter is employed. Applying the equations given at the end of this chapter, the reader can use this technique extended by directional (adaptive) forgetting. Application of this method is illustrated by several examples.

This chapter is divided into the following sections. Section 3.1 presents stochastic process models suitable for self-tuning controller design. The recursive least squares method is discussed in detail in Section 3.2, together with an algorithm for the advanced forgetting technique.

3.1 Stochastic Process Models

In order to create a model we seek a function f which describes the plant output behaviour $y(t)$ as a function of input variables, typically the controller output $u(t)$, and other measured variables which may affect the output, such as the disturbance variable $v(t)$. Hence we assume

$$y(t) = f\left[u(t), v(t), t\right] \qquad (3.1)$$

However, the output of an actual plant is rarely a deterministic function of the measurable input variables. Disturbance appears in the plant output, representing nonmeasurable influences on the process, variations in the operating point, in the raw materials composition, and so on. These influences, which are often very difficult to describe accurately, are included as random – stochastic – influences. The most generalized form of the model can then be characterized by the relation

$$y(t) = f\left[u(t), v(t), t\right] + n(t) \qquad (3.2)$$

where $n(t)$ is the term describing stochastic influences.

Discrete Model

While it is necessary to work with derivatives of measured signals when describing a continuous-time dynamic plant, it is considerably simpler to construct discrete models, which rely on signal values taken only at regular sampling periods T_0. The digital process computer, which is used as the control unit of the control loop, operates only in discrete time sequences $t_k = kT_0 (k = 0, 1, 2, ...)$. In the case of controlling a continuous-time technological process we consider a continuous-time control object and a discrete controller. For this control loop, in order to work properly, an interface between these differently operating dynamic systems is essential. Sample and hold units are used as the interface. The sample part samples the continuous-time signal in k-multiples of sampling periods T_0 to produce an output signal as an impulse sequence. The height of the impulses is equal to the value of the input signal at the sampling instant. For control of technological processes, zero-order hold is used almost exclusively to hold the impulse constant over the entire sampling period. Therefore, it is necessary to use a suitable mathematical description to express the dynamic behaviour of components of the control loop discretized in this way. One such description is an expression employing the $\mathcal{Z}$-transform. If $G(s)$ is the transfer function of a continuous-time dynamic system, then the following expression for the discrete transfer function with zero-order holder is valid

$$G(z) = (1 - z^{-1})\mathcal{Z}\left\{\mathcal{L}^{-1}\frac{G(s)}{s}\right\}_{T_0} \tag{3.3}$$

This step transfer function (3.3) is a rational polynomial function in the complex variable z. Note that the complex variable z has the meaning of the forward time-shift operator and z^{-1}, the backward time-shift operator, so

$$zy(k) = y(k + 1) \quad z^{-1}y(k) = y(k - 1)$$

The simple model structure, identification from measured data, suitability for the synthesis of the discrete control loop and for the description and expression of different types of stochastic processes including disturbance modelling are advantages of the discrete-time transfer function (3.3).

Predictor

It is possible to express a generalized discrete description of a dynamic system as a function of the values of previously measured variables, *i.e.*

$$y(k) = f[y(k - 1), y(k - 2), \ldots, y(k - na), u(k - 1), u(k - 2), \ldots, $$
$$u(k - nb), v(k - 1), v(k - 2), \ldots, v(k - nd), k] + n(k) \tag{3.4}$$

where $y(k)$ is the value of the output variable at the k-sample interval, *i.e.* at time $t = kT_0$ (T_0 is the sampling period which in Equation (3.4) is taken to be one). The problem is how to specify the stochastic term more precisely. Disturbance $n(k)$ can be modelled as a signal originating as a noise signal with known characteristics passing via a given filter. In the same way as the plant, the filter can be described in relation to delayed input and output variables. Thus we obtain

$$
\begin{aligned}
y(k) = f[&y(k-1), y(k-2), \ldots, y(k-na), u(k-1), u(k-2), \ldots, u(k-nb), \\
&v(k-1), \ldots, v(k-nd), e_s(k), e_s(k-1), e_s(k-2), \ldots, e_s(k-nc), k]
\end{aligned}
$$
$$(3.5)$$

where $e_s(k)$ is the random nonmeasurable component. If we limit ourselves to a linear function f, it is possible to obtain the familiar ARMAX model

$$
\begin{aligned}
y(k) = &- \sum_{i=1}^{na} a_i y(k-i) + \sum_{i=1}^{nb} b_i u(k-i) + \sum_{i=1}^{nd} d_i v(k-i) \\
&+ e_s(k) + \sum_{i=1}^{nc} c_i e_s(k-i)
\end{aligned}
$$
$$(3.6)$$

or in the probably better known form using the backward time-shift operator z^{-1}

$$
A\left(z^{-1}\right) y(k) = B\left(z^{-1}\right) u(k) + D\left(z^{-1}\right) v(k) + C\left(z^{-1}\right) e_s(k) \qquad (3.7)
$$

where the individual polynomials of equation (3.7) take the form

$$
\begin{aligned}
A\left(z^{-1}\right) &= 1 + a_1 z^{-1} + a_2 z^{-2} + \ldots + a_{na} z^{-na} \\
B\left(z^{-1}\right) &= b_1 z^{-1} + b_2 z^{-2} + \ldots + b_{nb} z^{-nb} \\
C\left(z^{-1}\right) &= 1 + c_1 z^{-1} + c_2 z^{-2} + \ldots + c_{nc} z^{-nc} \\
D\left(z^{-1}\right) &= d_1 z^{-1} + d_2 z^{-2} + \ldots + d_{nd} z^{-nd}
\end{aligned}
$$
$$(3.8)$$

However, the ARMAX model is not entirely suitable for adaptive control. If its parameters (coefficients of polynomials A, B, C, D) are the subject of identification using measured data we encounter the problem of identifying coefficients of the polynomial $C(z^{-1})$ because the fictitious noise $e_s(k)$ cannot be measured. Although there are identification procedures (the extended least squares method, for example) enabling $C(z^{-1})$ to be identified, their convergence is not guaranteed generally and usually is too slow. Therefore most adaptive controller designs are based on the regression (ARX) model, which describes the plant output according to the relation

$$
y(k) = - \sum_{i=1}^{na} a_i y(k-i) + \sum_{i=1}^{nb} b_i u(k-i) + \sum_{i=1}^{nd} d_i v(k-i) + e_s(k) \qquad (3.9)
$$

or

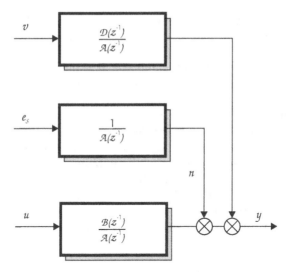

Figure 3.1. Block diagram of the regression model

$$A\left(z^{-1}\right) y(k) = B\left(z^{-1}\right) u(k) + D\left(z^{-1}\right) v(k) + e_s(k) \tag{3.10}$$

The block diagram of this model is given in Figure 3.1.

The ARX regression model is often written in the compact vector form

$$y(k) = \boldsymbol{\Theta}^T(k)\boldsymbol{\phi}(k-1) + e_s(k) \tag{3.11}$$

where

$$\boldsymbol{\Theta}^T(k) = [a_1, a_2, \ldots, a_{na}, b_1, b_2, \ldots, b_{nb}, d_1, d_2, \ldots, d_{nd}] \tag{3.12}$$

is the vector of parameters of the model under examination and

$$\begin{aligned}
\boldsymbol{\phi}^T(k-1) = [&-y(k-1), -y(k-2), \ldots, -y\left(k-na\right), \\
&u(k-1), u\left(k-2\right), \ldots, u\left(k-nb\right), \\
&v(k-1), v\left(k-2\right), \ldots, v\left(k-nd\right)]
\end{aligned} \tag{3.13}$$

is the data vector, so-called regressor.

Note:
The quality of the model largely depends on the sampling period and order of the regression model. In modelling and simulation, it is necessary to fulfil the rule that a discrete open-loop must maintain at least one time-delay interval even if the delay is not included in the continuous-time transfer functions. Otherwise algebraic loops occur when the loop is closed, which is unrealistic. In the literature it is possible to find cases where the polynomial $B(z^{-1})$ is considered either with the term b_0 or without even for plants with no time

delay. If $b_0 \neq 0$ is considered, then controller output $u(k)$ affects $y(k)$. If this holds, then the controller may not use $y(k)$ to generate $u(k)$. On the other hand, if a controller is used such that $u(k) = f[y(k)]$ for plants which have no time delay, then $b_0 = 0$ should be fulfilled. It is difficult to choose between these two possibilities but it is always necessary to decide which convention is to be used in each case. It may happen that an extra time delay occurs in the loop. Relations (3.6)–(3.9) are used in Chapters 4 and 5, which discuss self-tuning PID controllers and controllers derived from the algebraic theory of control. In Chapter 6 where we discuss STCs based on the minimization of the LQ (Linear Quadratic) criteria, the parameter vector is expanded to include the coefficient b_0 and consequently, the data vector includes the variable $u(k)$. As a result, the variable $u(k)$ appears in functions (3.4) and (3.5), and also in equations (3.9) and (3.13).

The quality of the regression model used is primarily judged by the prediction error, *i.e.* the deviation

$$\hat{e}(k) = y(k) - \hat{y}(k) \tag{3.14}$$

where $\hat{y}(k)$ is calculated according to (3.9) with $e_s(k) = 0$. The prediction error plays a key role in identifying the parameters for a regression model derived from measured data. It is also important for selecting the structure (order) of the regression model and a suitable sampling period. It must be emphasized that for a good model, the prediction error must not only be small but must also represent white (uncorrelated) noise with a zero mean value.

The quality of the model is also judged by the purpose for which it is used. The crucial model characteristics are those used for controller design. For example, in the Ziegler–Nichols PID design method, the dominant characteristic is the ultimate gain of the plant. When using the pole assignment design method, we require the model to mirror the placement of zeros and poles faithfully. For the LQ approach, the stochastic term is also required to represent the disturbance under consideration.

3.2 Process Identification

In adaptive control the task of identification is just as important as the role of control synthesis. Identification for adaptive control has, of course, its own specification, which for most cases involves estimation of parameters for the ARX regression model using the least squares method. Here, we can explore reasons for this. When identifying a given plant, we should follow this procedure:

1. Preparation of the identification experiment. A choice of the most suitable input (exciting) signal, a trade-off between the theoretical optimal excitement and that applied, with respect to the technology used. The process

of identification can be observed, interrupted, and the input signal can be altered.

2. The data gathered during the experiment can be stored and subsequently processed using various methods with different models, filtered, *etc.*
3. The model parameters obtained can be tested using other sampled data.
4. The identification experiment can be repeated, possibly with the knowledge gained from previous experiments.
5. Conditions for the lack of bias of the estimates can be tested or verified.

However, when performing identification for adaptive control, it is necessary to meet the following conditions:

- The data (inputs) are generated by a feedback controller.
- The aim of the controller is to compensate for disturbances and to stabilize the process. These circumstances make it more difficult to identify process parameters.
- The identification process for adaptive control takes an infinitely long time. It is, therefore, not possible to assume constant parameter estimates. Methods to estimate time-varying parameters are essential.
- The identification must be functional with various operating conditions of the plant (at relative steady-state, with disturbances, and during transitions between different states).
- The structure of the identified model (order) cannot usually be changed while running.
- The identification algorithm must be numerically reliable and sufficiently fast.

We see from this that the conditions governing adaptive control are not always ideal for identification. The conditions for obtaining unbiased estimates cannot usually be tested, they can only be assumed in these cases. If the assumptions are not met, adaptive control can get into difficulties.

3.2.1 Typical Identification Problems in Adaptive Control

In this section we demonstrate several specific problems in identification for controller design.

a priori information.

When using the certainty equivalence principle, the model parameters must approach the true values right from the start of control. This means that as the adaptive algorithm begins to operate, identification must be run from suitable start-up conditions – the result of the best possible *a priori* information. Their role in identification is often underestimated. In a typical identification experiment their role is not, in fact, so significant since we are only interested in the results of identification at the end of the experiment when sufficient data have been analysed. In adaptive control, it is important to include every possible piece of information in the start-up conditions for identification, in particular for the following reasons:

- the parameter estimates must represent the plant right from the start of the identification process to prevent the designed controller from performing inadequately;
- data obtained as the controller is operating are not always sufficiently informative and in this case *a priori* information provides the minimum safe information.

The start-up conditions for the most commonly used identification methods are represented by initial parameter estimates and their covariance matrix. Although most users understand the importance of the initial parameter estimates and with a certain amount of effort are usually able to assign realistic values using their technical expertise, the importance of the covariance matrix is often neglected and it is difficult to estimate. The fictitious data method [40] has proved to be a viable and relatively simple way to obtain the start-up conditions for identification, including more or less all *a priori* information. It works by means of a model (which can be very simple) representing the characteristic under analysis to generate the data.

For example, if we know that the plant gain equals "g", then the corresponding output value to any $u(t)$ input value will be $y(t) = gu(t)$. If we know one point of the frequency characteristic, then it is possible to obtain more data so that the input will be a sine wave with a given frequency and the output will be a shifted sine wave with the amplitude corresponding to the absolute value of the frequency characteristic of the plant at that point and the shift corresponding to the phase.

Similarly, data can be generated for some other requested information. In less complex examples data can be contrived, for more complicated cases data can be obtained using simulation.

By processing this data in the same way as if they were real, it is possible to obtain the start-up estimates and the covariance matrix. The problem is that this data cannot be processed using the common approaches (for example the least squares method). We have to take into account that individual components of the *a priori* information may be partially conflicting, but in any case, this information can only be considered to have a certain probability. It may happen that the use of a great deal of data on specific information (for example gain) leads to the information becoming so fixed in the estimates that even a large amount of real data cannot change it.

In practice, it turns out that even a very small amount of *a priori* information can have a positive effect on the start-up of adaptive control if it is correctly introduced. *a priori* information is also important for determining the model structure when preparing adaptive control (see Section 6.5).

Monitoring time-variable parameters.

As was noted earlier, the assumption of constant parameters is unacceptable when considering adaptive controllers, mainly for the following reasons:

- the long period of operation of the controller;
- change in the parameters of a linearized model together with alteration of the plant's operating point.

The issue of estimating time-variable parameters is detailed and tested in [34] and [41]. When there is no information on the character of changes in the parameters, the problem can be solved using a forgetting technique. The best known is the exponential forgetting factor method where the influence of old data on the parameter estimates and their covariance matrix decreases exponentially. A serious drawback of this technique in an adaptive mode is the loss of information when the process is stabilized so that the data cannot provide enough information on it. This situation can be resolved by switching off the identification process, using the variable forgetting factor or other forms of forgetting (directional, regularized) which have the ability to alter the amount of information forgotten according to the character of the data.

3.2.2 Identification Algorithms

It is not necessary to differentiate between coefficients a_i and b_i in order to identify the parameters of a regressive model; it is possible to work with the vector of unknown parameters $\Theta(k)$ (3.12) and the regression vector $\phi(k-1)$ (3.13).

For control purposes using a self-tuning controller we are interested only in those methods of experimental identification which can be performed in real time. Recursive procedures are the most suitable for parameter estimation in real time when the estimate in the discrete-time step k is obtained using new data to correct an earlier estimate $\hat{\Theta}(k-1)$ in time $k-1$. The most common recursive procedures [42, 43] to estimate the parameters of an ARX model are the following:

- the recursive least squares method;
- the recursive instrumental variable method;
- the stochastic approximation;

and to estimate the parameters of an ARMAX model:

- the extended recursive least squares method;
- the recursive maximum likelihood method.

The least squares method described by Ljung and Söderström [44] and Strejc [45] has given the best results in estimation of the ARX model parameters and it will be used in the identification procedure of all the self-tuning controllers presented in this monograph.

3.2.3 Principle of the Least Squares Method

The least squares method is one of the methods of regression analysis suitable for examining static and dynamic relations between variables of the plant under consideration. Consider a single-input/single-output (SISO) stochastic process described by the ARX model (3.11) and for the parameter vector (3.12) and regression vector (3.13) assume that $na = nb = n, nd = 0$, *i.e.* their dimension is $nz = 2n$

$$\Theta^T(k) = [a_1, a_2, \ldots, a_n, b_1, b_2, \ldots, b_n] \tag{3.15}$$

$$\phi^T(k-1) = [-y(k-1), -y(k-2), \ldots, -y(k-n), u(k-1), u(k-2), \ldots, u(k-n)] \tag{3.16}$$

Then the generation of output signal $y(k)$ at individual time-moments can be expressed using a matrix equation

$$\mathbf{y} = \mathbf{F\Theta} + \mathbf{e} \tag{3.17}$$

where the matrix $\mathbf{F}$ of dimension $(N - n, 2n)$ and vectors $\mathbf{y}, \mathbf{e}$ of dimension $(N - n)$ take the form

$$\mathbf{y}^T = [y(n+1), y(n+2), \ldots, y(N)] \tag{3.18}$$

$$\mathbf{e}^T = [e_s(n+1), e_s(n+2), \ldots, e_s(N)] \tag{3.19}$$

$$\mathbf{F} = \begin{bmatrix} -y(n) & -y(n-1) & \ldots & -y(1) & u(n) & \ldots & u(1) \\ -y(n+1) & -y(n) & \ldots & -y(2) & u(n+1) & \ldots & u(2) \\ \vdots & & & & & & \vdots \\ -y(N-1) & -y(N-2) & \ldots & -y(N-n) & u(N-1) & \ldots & u(N-n) \end{bmatrix} \tag{3.20}$$

N is the number of samples of measured input and output data. From Equation (3.17) an error can be determined as

$$\mathbf{e} = \mathbf{y} - \mathbf{F\Theta} \tag{3.21}$$

then, the following criterion is defined

$$J = \mathbf{e}^T\mathbf{e} = (\mathbf{y} - \mathbf{F\Theta})^T(\mathbf{y} - \mathbf{F\Theta}) \tag{3.22}$$

the minimum of which can be obtained by differentiating (3.22) with respect to Θ and setting this equal to zero, *i.e.*

$$\left. \frac{\partial J}{\partial \Theta} \right|_{\Theta = \hat{\Theta}} = 0 \tag{3.23}$$

By solving Equation (3.23) it is possible to obtain a basic matrix relation to estimate model parameters using the least squares method

$$\hat{\Theta} = \left(\mathbf{F}^T \mathbf{F} \right)^{-1} \mathbf{F}^T \mathbf{y} \qquad (3.24)$$

Equation (3.24) serves for a one-off calculation of the parameter estimates of the process model using N samples of measured data. This calculation places fairly high demands on the computer memory which must be large enough to store all the measured data.

Example 3.1. Assume a plant with transfer function $G(z) = \frac{0.1997z^{-1}}{1-0.8669z^{-1}}$. Assume the input to the plant to be a random signal $u(k)$ with Gaussian distribution, zero expected value and a unity variance. The response $y(k)$ of the system was recorded in the form of Table 3.1 and Figure 3.2. Measurement noise was simulated by a random signal with standard deviation 0.001 affecting the output from the plant. For identification, assume a discrete ARX model in the form $y(k) = -a_1 y(k-1) + b_1 u(k-1) + e_s(k)$ and compute the vector of process model parameter estimates $\hat{\Theta}^T = \left[\hat{a}_1, \hat{b}_1 \right]$ and error vector $\mathbf{e}$ using the least squares method. Compare the calculated vector of parameter estimates $\hat{\Theta}$ at time $k = 8$ for sample period $T_0 = 1$ s with the parameters a_1, b_1 of the discrete transfer function $G(z) = \frac{0.1997z^{-1}}{1-0.8669z^{-1}}$.

Table 3.1. Record of input and output variables of the example system

k	1	2	3	4	5	6	7	8
$u(k)$	-0.6918	0.8580	1.2540	-1.5937	-1.4410	0.5711	-0.3999	0.6900
$y(k)$	0.0071	-0.1252	0.0583	0.3070	0.0744	-0.3420	-0.1839	-0.2539

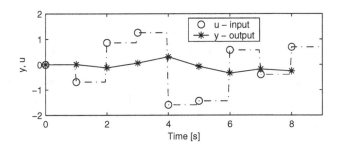

Figure 3.2. Example 3.1: record of input and output variables

The number of measured pairs of input and output values is $N = 8$, model order $n = 1$. Data matrix $\mathbf{F}$ of dimension $(7, 2)$ and data vector $\mathbf{y}$ of dimension $(7, 1)$ are in the form (see (3.18) and (3.20))

$$\mathbf{F} = \begin{bmatrix} -0.0071 & -0.6918 \\ 0.1252 & 0.8580 \\ -0.0583 & 1.2540 \\ -0.3070 & -1.5937 \\ 0.0744 & -1.4410 \\ 0.3420 & 0.5711 \\ 0.1839 & -0.3999 \end{bmatrix}$$

$$\mathbf{y}^T = [-0.1252, 0.0583, 0.3070, -0.0744, -0.3420, -0.1839, -0.2539]$$

By substitution into Equation (3.24) the parameter estimates vector is given by $\hat{\Theta}^T = [-0.8593, 0.2023]$ and using Equation (3.21) it is possible to obtain the error vector.

$$\mathbf{e}^T = [0.0087, -0.0077, 0.0032, -0.0158, 0.0135, -0.0055, -0.0150]$$

3.2.4 Recursive Identification Using the Least Squares Method

Equation (3.24) cannot be used to calculate the parameter estimates of the process model for self-tuning controllers; it is necessary to use its recursive version, which can perform the identification in real time. Here, newly measured values are only used to correct the original estimates. This reduces the complexity of the calculation and thus the demands placed on the computer technology used. Recursive algorithms allow one to monitor changes in the characteristics (parameters) of the process in real time and therefore form the basis for self-tuning controllers.

Let a linear SISO stochastic model be described by the ARX model expressed in the form (3.11). Further, the nonmeasurable random component $e_s(k)$ is a sequence of mutually uncorrelated random signals uncorrelated also with the process input and output. Assume furthermore that the random variable has an expected value equal to zero and constant variance.

Now the task is to estimate recursively the unknown parameters Θ of model (3.11) on the basis of inputs and outputs at time k, $\{u(i), y(i), i = k, k-1, k-2, \ldots, 0\}$. Therefore, the unknown vector parameter Θ of dimension $nz = 2n$ can be found when minimizing the criterion

$$J_k(\Theta) = \sum_{i=0}^{k} e_s^2(i) \tag{3.25}$$

where

$$e_s(i) = y(i) - \Theta^T \phi(i) = \begin{bmatrix} 1 & -\Theta^T \end{bmatrix} \begin{bmatrix} y(i) \\ \phi(i) \end{bmatrix} \tag{3.26}$$

If we require the algorithm to be able to monitor slow changes in the parameters of the identified process, this can be achieved using the technique of exponential forgetting. Then, it is necessary to minimize the modified criterion

$$J_k(\boldsymbol{\Theta}) = \sum_{i=0}^{k} \varphi^{2(k-i)} e_s^2(i) \tag{3.27}$$

where $0 < \varphi^2 \leq 1$ is the exponential forgetting factor. Substitution of Equation (3.26) into criterion (3.27) yields

$$J_k(\boldsymbol{\Theta}) = \left[1 - \boldsymbol{\Theta}^T\right] \mathbf{V}(k) \begin{bmatrix} 1 \\ -\boldsymbol{\Theta} \end{bmatrix} \tag{3.28}$$

The symmetrical square matrix $\mathbf{V}(k)$ of type $(nz + 1, nz + 1)$, which is assumed to be positively definite, is given as

$$\mathbf{V}(k) = \sum_{i=k_0}^{k} \varphi^{2(k-i)} \mathbf{d}(i)\mathbf{d}^T(i); \quad \mathbf{d}(i) = [y(i)\phi(i)]^T \tag{3.29}$$

and can be calculated recursively

$$\mathbf{V}(k) = \varphi^2 V(k-1) + \mathbf{d}(k)\mathbf{d}^T(k) \tag{3.30}$$

Now, it is clear that minimization of criterion (3.27) leads to the minimization of quadratic form (3.28) from the point of view of the parameter vector $\boldsymbol{\Theta}$. Minimization of (3.28) leads to equations containing $\mathbf{V}^{-1}$.

Positive semi-definiteness is a necessary characteristic of the matrix $\mathbf{V}^{-1}(k)$ because it ensures the non-negativity of the minimized function, and consequently, the existence of a finite minimizing argument $\hat{\boldsymbol{\Theta}}(k)$.

In numerically adverse conditions (matrix $\mathbf{V}^{-1}(k)$ is almost singular), which is common during operation of self-tuning controllers, it is necessary to use the version of the least squares method where the theoretically assumed positive semi-definiteness of $\mathbf{V}(k)$ holds also numerically. Otherwise the identification may collapse numerically. These numerical difficulties motivated the development of filters to prevent the numeric collapse of the algorithm. Recently, the so-called square root filter REFIL derived by Peterka [46] was successfully used. The basic idea of this digital filtration technique is to replace recursive relations for the calculation of the symmetric matrix (which must be positively semi-definite) like Equation (3.30) with a recursive calculation of the square root of this matrix. The so-called Cholesky square root inversion matrix $\mathbf{V}(k)$ has been shown to work well.

$$\mathbf{G}(k) = \left[\mathbf{V}^{-1}(k)\right]^{\frac{1}{2}} \tag{3.31}$$

The Cholesky square root positively semi-definite matrix $\mathbf{V}^{-1}(k)$ is defined as the lower triangular matrix (3.30) which has non-negative elements on the main diagonal and fulfils relation

$$\mathbf{V}^{-1}(k) = \mathbf{G}(k)\mathbf{G}^T(k) \tag{3.32}$$

whereas the transposition of the Cholesky square root $\mathbf{G}^T(k)$ is the upper triangular matrix. The numerical advantage of square root filtration lies in the fact that if we recursively calculate square root $\mathbf{V}^{-1}(k)$ instead of matrix $\mathbf{G}(k)$, then whatever real matrix $\mathbf{G}(k)$ is, the product (3.32) is always a positive semi-definite matrix.

Here, an alternative filter LDFIL will be used which retains the necessary numeric characteristics of the REFIL filter but does not require the square roots of diagonal elements to be determined and also saves on the number of multiplications [47, 48].

Let us consider the factorization of matrix $\mathbf{V}(k)$ in the form

$$\mathbf{V}^{-1}(k) = \mathbf{L}(k)\mathbf{D}(k)\mathbf{L}^T(k) \tag{3.33}$$

where $\mathbf{D}(k)$ is a diagonal matrix (with positive elements) and $\mathbf{L}(k)$ a lower triangular matrix with a unit diagonal (both matrices are square with dimension $nz + 1$). This kind of factorization certainly exists for a positively definite (regular and semi-definite) matrix. If the matrices $\mathbf{D}$ and $\mathbf{L}$ are separated into blocks (for now the discrete time k is left out for the sake of clarity)

$$\mathbf{D} = \begin{bmatrix} \mathbf{D}_y & \mathbf{0} \\ \mathbf{0} & \mathbf{D}_z \end{bmatrix} \qquad \mathbf{L} = \begin{bmatrix} \mathbf{1} & \mathbf{0} \\ \mathbf{L}_{zy} & \mathbf{L}_z \end{bmatrix} \tag{3.34}$$

then, it is possible to rewrite criterion (3.28) as

$$J_k(\boldsymbol{\Theta}) = \begin{bmatrix} 1 \\ -\boldsymbol{\Theta} \end{bmatrix}^T (\mathbf{L}^{-1})^T \mathbf{D}^{-1} \mathbf{L}^{-1} \begin{bmatrix} 1 \\ -\boldsymbol{\Theta} \end{bmatrix} \tag{3.35}$$

Since

$$\mathbf{L}^{-1} = \begin{bmatrix} 1 & 0 \\ -\mathbf{L}_z^{-1}\mathbf{L}_{zy} & \mathbf{L}_z^{-1} \end{bmatrix} \tag{3.36}$$

is valid for the inversion of triangular matrix $\mathbf{L}$, the criterion (3.35) can be rewritten as

$$
\begin{aligned}
J_k(\boldsymbol{\Theta}) &= \begin{bmatrix} 1 \\ -\boldsymbol{\Theta} \end{bmatrix}^T \begin{bmatrix} 1 & -\mathbf{L}_z^{-1}\mathbf{L}_{zy} \\ 0 & \mathbf{L}_z^{-1} \end{bmatrix} \begin{bmatrix} \mathbf{D}_y^{-1} & 0 \\ 0 & \mathbf{D}_z^{-1} \end{bmatrix} \begin{bmatrix} 1 & 0 \\ -\mathbf{L}_z^{-1}\mathbf{L}_{zy} & \mathbf{L}_z^{-1} \end{bmatrix} \begin{bmatrix} 1 \\ -\boldsymbol{\Theta} \end{bmatrix} \\
&= \mathbf{D}_y^{-1} + \begin{bmatrix} -\boldsymbol{\Theta} & -\mathbf{L}_{zy} \end{bmatrix}^T \begin{bmatrix} \mathbf{L}_z^{-1} \end{bmatrix}^T \mathbf{D}_z^{-1}\mathbf{L}_z^{-1} \begin{bmatrix} -\boldsymbol{\Theta} & -\mathbf{L}_{zy} \end{bmatrix}
\end{aligned}
\tag{3.37}
$$

It is clear from the form of criteria (3.37) that only the second non-negative addend depends on model parameters Θ on the right-hand side. The absolute minimum is therefore obtained for

$$\hat{\Theta}(k) = -\mathbf{L}_{zy}(k) \tag{3.38}$$

and the value of this minimum is

$$J_k(\hat{\Theta}) = \mathbf{D}_y^{-1}(k) \tag{3.39}$$

The solution to this problem is thus contained in the factorization of (3.33), which can be illustrated as

$$\mathbf{L}(k) = \begin{bmatrix} 1 & 0 \\ -\hat{\Theta}(k) & \mathbf{L}_z(k) \end{bmatrix}; \quad \mathbf{D}(k) = \begin{bmatrix} [\min J_k(\hat{\Theta})]^{-1} & 0 \\ 0 & \mathbf{D}_z(k) \end{bmatrix} \tag{3.40}$$

In conclusion to this section the algorithm of the recursive least squares method is given, regardless of the numerical aspects mentioned above, extended to include the technique of directional (adaptive) forgetting [49, 50].

The vector of parameter estimates is updated according to the recursive relation

$$\hat{\Theta}(k) = \hat{\Theta}(k-1) + \frac{\mathbf{C}(k)\phi(k-1)}{1 + \xi(k)}\hat{e}(k) \tag{3.41}$$

where

$$\xi(k) = \phi^T(k-1)\mathbf{C}(k)\phi(k-1) \tag{3.42}$$

is an auxiliary scalar and

$$\hat{e}(k) = y(k) - \hat{\Theta}^T(k-1)\phi(k-1) \tag{3.43}$$

is the prediction error. If $\xi(k) > 0$, the square covariance matrix with dimension nz is updated by relation

$$\mathbf{C}(k) = \mathbf{C}(k-1) - \frac{\mathbf{C}(k-1)\phi(k-1)\phi^T(k-1)\mathbf{C}(k-1)}{\varepsilon^{-1}(k) + \xi(k)} \tag{3.44}$$

where

$$\varepsilon(k) = \varphi(k) - \frac{1 - \varphi(k)}{\xi(k-1)} \tag{3.45}$$

If $\xi(k) = 0$, then

$$\mathbf{C}(k) = \mathbf{C}(k-1) \tag{3.46}$$

The value of adaptive directional forgetting $\varphi(k)$ is then calculated for each sampling period as

$$[\varphi(k)]^{-1} = 1 + (1 + \rho)[\ln(1 + \xi(k))] + \left[\frac{(v(k) + 1)\eta(k)}{1 + \xi(k) + \eta(k)} - 1\right]\frac{\xi(k)}{1 + \xi(k)} \tag{3.47}$$

Listing 3.1. Recursive least squares method with directional forgetting for a second-order model – identdf.m

```
% Recursive least squares method with directional forgetting
% fi     - directional forgetting factor
% theta  - vector of the parameter estimates
% d      - regression vector
% c      - covariance matrix
% la     - parameter lambda
% ny     - parameter eta
% ep     - prediction error
% ks     - auxiliary scalar ksi
% eps    - auxiliary parameter
% te     - parameter theta
% up     - previous controller output
% y      - current process output
% ro     - parameter ro

function
[fi,theta,d,c,la,ny,ep,te,ks,pp]=idendf(fi,theta,d,c,la,ny,up,y,ro)

% Cyclic date substitution in regression vector (1)
d(4)=d(3);
d(3)=up;

% Update of parameter estimates
ep=y-theta'*d;      %prediction error
ks=d'*c*d;
pp=(c*d/(1+ks))*ep;
theta=theta+pp;

% Update of identification variables
if ks>0
    eps=fi-(1-fi)/ks;
    c=c-c*d*d'*c/(inv(eps)+ks);
end
la=fi*(la+ep*ep/(1+ks));
ny=fi*(ny+1);
te=ep*ep/la;
fi=1/(1+(1+ro)*(log(1+ks-ks/(1+ks)+ks*(ny+1)*
    te/(1+ks+te)/(1+ks))));
% Cyclic date substitution in regression vector (2)
d(2)=d(1);
d(1)=-y;
```

where

$$\eta(k) = \frac{\hat{e}^2(k)}{\lambda(k)}$$

$$v(k) = \varphi(k)\left[(v(k-1)+1\right]$$

$$\lambda(k) = \varphi(k) \left[\lambda(k-1) + \frac{\hat{e}^2(k)}{1 + \xi(k)} \right] \tag{3.48}$$

are auxiliary variables. Although the question of *a priori* information in the selection of start-up conditions has been discussed in Section 3.2.1, it proved better to choose the following conditions for the start of the algorithm: elements of the main diagonal of the covariance matrix $C_{ii}(0) = 10^3$, start value for the directional forgetting factor $\varphi(0) = 1, \lambda(0) = 0.001, v(0) = 10^{-6}$, $\rho = 0.99$. The initial estimates for the vector $\hat{\Theta}(0)$ are chosen according to *a priori* information and this selection has caused no problems in the majority of simulation and laboratory tests on self-tuning controllers.

The relations given above can directly be programmed as an M-function in the MATLAB® system without taking the numeric aspects into account, *i.e.* without the use of numeric filters.

The algorithm in Listing 3.1 is the implementation of the recursive least squares method with directional forgetting factor for a second-order model and in Listing 3.2 is an example for the initial procedure.

Listing 3.2. Initialization of identification for a second-order model – inide.m

```
% Initialization of identification
function [d,theta,c,ro,fi,la,ny,u]=inide
d=zeros(4,1);                % regression vector
theta=[0.1; 0.2; 0.1; 0.2]; % parameter estimates
c=1000*eye(4);              % covariance matrix
% Initialization of auxiliary parameters
ro=0.99; fi=1; la=0.001; ny=0.000001; u=0;
```

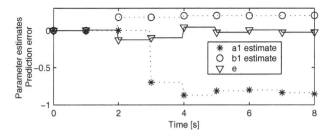

Figure 3.3. Example 3.2. Parameter estimates and prediction error

Example 3.2. Use the data listed in Table 3.1 of Example 3.1 for recursive identification of the process employing the least squares method with the first-order regression model $y(k) = -a_1 y(k-1) + b_1 u(k-1) + e_s(k)$. For parameter estimates, choose the forgetting factor $\varphi(k) = 1$. Plot a graph showing evolution of the parameter estimates and prediction error $\hat{e}(k)$. Compare the results with those obtained in Example 3.1.

It is clear that according to (3.10)–(3.12) it is possible to describe the regression model in the vector form:

$$y(k) = \mathbf{\Theta}^T(k)\boldsymbol{\phi}(k-1) + e_s(k) = [a_1 \; b_1] \begin{bmatrix} -y(k-1) \\ u(k-1) \end{bmatrix} + e_s(k)$$

Equations (3.41)–(3.48) were used for recursive computation of the parameter estimates with forgetting factor $\varphi(k) = 1$. The procedure *identbasic.m* (see MATLAB® Listing 3.3) was used for solving this example.

Figure 3.3 shows the process model parameter estimates and the prediction error. The converged vector of parameter estimates at step $k = 8$ is $\hat{\mathbf{\Theta}}^T(8) = [-0.8556, 0.2020]$. The results obtained match those from Example 3.1.

Example 3.3. Assume the block scheme of an ARX model as shown in Figure 3.1 neglecting the measurable error variable, *i.e.* $v(t) = 0$. The process is modelled by the continuous-time transfer function

$$G(s) = \frac{Y(s)}{U(s)} = \frac{B(s)}{A(s)} = \frac{2.5(0.8s+1)}{(4.2s+1)(0.5s+1)}$$

Identify this system by a recursive least squares method and determine discrete model parameter estimates for the sampling period $T_0 = 0.25$ s. For excitation of the model by signal $u(t)$ use a random signal generator with variance limits defined from 0 to 1. Generate the noise $e_s(t)$ by MATLAB® function *rand* with variance $10e^{-5}$. Display the evolution of model parameter estimates and prediction error for the time interval $\langle 0; 5 \rangle$ s.

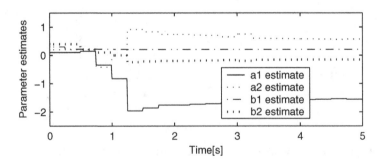

Figure 3.4. Example 3.3. Evolution of parameter estimates

Listing 3.3. Basic recursive least squares algorithm – identbasic.m

```
% Basic recursive least squares method
% N - number of identification steps
% c - covariance matrix
% d - regression vector
% theta - vector of the parameter estimates
% ep(k)- prediction error
% eps - auxiliary parameter
% y(k)  - process output
% u(k)  - controller output

% input variables
u=[-0.6918 0.858 1.254 -1.5937 -1.441 0.5711 -0.3999 0.69];
% output variables
y=[0.0071 -0.1252 0.0583 0.307 -0.0744 -0.342 -0.1839 -0.2539];
N=8;
theta= [0 0]';  % initial vector of parameter estimates
c = 1000*eye(2);  % initial covariance matrix
for k=2:1:N
    d = [ -y(k-1)  u(k-1) ]'; % new data vector
    ep(k)= y(k)-theta'*d;
    eps = d'*c*d;
    theta = theta + (c*d*ep(k)) / (1+eps);  % theta update
    c = c-(c*d*d'*c)/(1+eps); % new covariance matrix
    a1(k)=theta(1);
    b1(k)=theta(2);
end
figure;
plot(a1,'k');
hold on;
plot(b1,'b');
plot(ep,'r');
xlabel('time steps')
ylabel('theta, e')
```

The parameter estimates evolution is shown in Figure 3.4 and the prediction error in Figure 3.5. The parameter estimates at time 5 s have the following values:

$$\hat{a}_1(5) = -1.5557; \quad \hat{a}_2(5) = 0.5781; \quad \hat{b}_1(5) = 0.2074; \quad \hat{b}_2(5) = -0.1547.$$

It is possible to compare the results obtained by recursive identification with the discrete transfer function of the plant for the sampling period $T_0 = 0.25$ s, which is given as

$$G(z) = \frac{0.2125z^{-1} - 0.1557z^{-2}}{1 - 1.5487z^{-1} + 0.5715z^{-2}}$$

Listing 3.4. Function for identification of an n-order model – identnorder.m

```
% Recursive least squares method for an n-order model
% theta - vector of the parameter estimates
% d    - regression vector
% c    - covariance matrix
% ep   - prediction error
% y      - new process output
% u      - new controller output
% na     - order of the denominator
% nb     - order of the numerator
function [theta,d,c,ep]=identnorder(na,nb,theta,d,c,u,y)
  d = [ y d(1:na-1) u  d(na+1:na+nb-1) ]'; % new data vector
  ep= y-theta'*d;
  eps = d'*c*d;
  theta = theta + (c*d*ep) / (1+eps);   % theta update
  c = c-(c*d*d'*c)/(1+eps); % new covariance matrix
end;
```

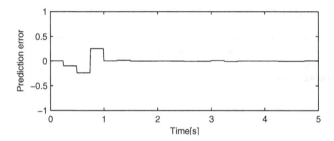

Figure 3.5. Example 3.3. Evolution of prediction error

Example 3.4. Assume the block scheme of the ARX model as shown in Figure 3.1, with the following difference: set the measurable error variable to $v(t) = \sin t$ and the polynomial $D(s) = 0.6s + 2$. Choose the same conditions for the experiment as in Example 3.3 but implement the identification in the following way:

1. In the first case, use the discrete regression model structure (3.8) for the identification. Neglect the measurable error variable $v(t)$, so the parameter estimates d_1, d_2 need not be computed (but you still have to consider that the input of the model is affected by errors).
2. In the second case, use the discrete regression model structure (3.8) for the identification and consider the measurable error variable $v(t)$, so the parameter estimates d_1, d_2 have to be computed.

Plot graphs of model parameter estimates and the prediction error for the time interval $\langle 0; 10 \rangle$ s. Compare the prediction error in both experiments according to the quadratic criterion in the form

$$S_e = \frac{1}{N} \sum_{k=1}^{N} [\hat{e}(k)]^2$$

where N is the number of identification steps.

Solution (a). The parameter estimates evolution and the prediction error are shown in Figure 3.6. The estimated transfer function of the process is
$G(z) = \frac{0.2147z^{-1} - 0.2344z^{-2}}{1 - 1.9330z^{-1} + 0.9428z^{-2}}$.

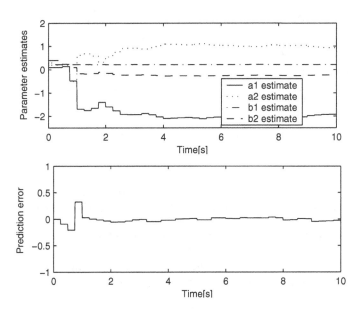

Figure 3.6. Example 3.4. Evolution of parameter estimates and prediction error – solution (a)

Criterion: $S_{ea} = \frac{1}{N} \sum_{k=1}^{N} [\hat{e}(k)]^2 = 0.0015$

Solution (b). The parameter estimates evolution and the prediction error are shown in Figure 3.7. The estimated transfer function of the process is
$G(z) = \frac{0.2129z^{-1} - 0.1546z^{-2}}{1 - 1.5439z^{-1} + 0.5670z^{-2}}$.

Criterion: $S_{eb} = \frac{1}{N} \sum_{k=1}^{N} [\hat{e}(k)]^2 = 0.0015$

In both cases, the prediction error converges to zero and values of the criteria are the same, however, in *solution (b)* the estimated parameters converge to the correct values of the process. In *solution (a)* the sine wave is identified as part of the plant and the estimated transfer function is close to the stabil-

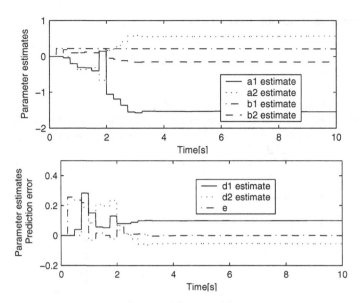

Figure 3.7. Example 3.4. Evolution of parameter estimates and prediction error – solution (b)

ity bound. Therefore, it is obvious that inclusion of parameters d_1, d_2 into the ARX model results in better accuracy of the identification experiment.

Example 3.5. Figure 3.8 shows a scheme for the laboratory model CE 108 – coupled drives apparatus (made by TecQuipment Ltd., Nottingham, United Kingdom). This model can be used for design and laboratory testing of control algorithms in real-time conditions. With this model, it is possible to design

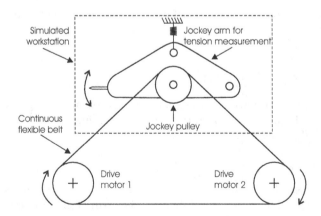

Figure 3.8. Example 3.5. Schematic diagram of the laboratory model CE 108

controllers for regulating the required speed and tension of the material on machines equipped with spools or cylinders. Examples of such devices are machines used for the production of fibres, paper, plastic foil, pipes, cables *etc.* In these technologies, the material passes a workstation which measures its speed and tension (the output variables of the process). These two variables are dependent on each other and can be controlled by changing the speed of motors placed in front and beyond the workstation (input variables of the process). On the laboratory model, the measurement of these variables is done by leading a flexible belt through three pulleys. The angular velocities of the two lower pulleys correspond to the rotation speeds of the motors. The third pulley is placed on a moving arm ("jockey arm") and plays the role of a workstation, measuring the speed and tension of the belt. Two drive motors' control the speed and tension of the rubber belt. The drive motors rotation speed range is 0–3000 revolutions/minute, which corresponds to the input power voltage 0–10 V. The tape tension measurement is done indirectly by the change of angle of the jockey arm ±10°, which corresponds to the output voltage of ±10 V. The amplifiers of the drive motors are bi-directional and allow control of the motor rotation in both directions.

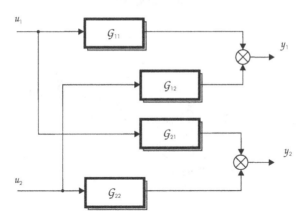

Figure 3.9. Example 3.5. Block diagram of the TITO system

From the description above it is clear that we are dealing with a two-input/two-output (TITO) system which can be described by the schema shown in Figure 3.9. The transfer matrix of the TITO system, which describes the input/output relations, can be written in the form:

$$\mathbf{G}(z) = \frac{\mathbf{Y}(z)}{\mathbf{U}(z)} = \begin{bmatrix} G_{11}(z) \; G_{12}(z) \\ G_{21}(z) \; G_{22}(z) \end{bmatrix}$$

where $\mathbf{U}(z)$ is the vector of input variables (voltage inputs to the motors) and $\mathbf{Y}(z)$ is the vector of controlled variables (tension and the tape speed).

The laboratory model shows quite nonlinear behaviour with strong interactions between the individual variables. The static characteristics of the model were measured to determine the system linearity ranges. All the characteristics show nonlinear behaviour, namely the belt tension characteristic is nonlinear over the whole range due to belt oscillations. The static characteristics are shown in Figure 3.10. The variable y_1 denotes speed and variable y_2 the tension of the belt. The variables u_1 and u_2 are the voltage inputs of the left and right drive motors.

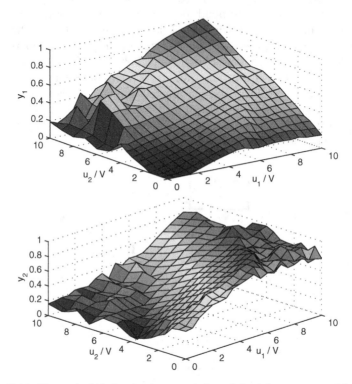

Figure 3.10. Example 3.5. Static characteristics of the laboratory model CE 108

Because the scope of this book is limited to single-input/single-output adaptive control systems, we show an example of experimental recursive identification of submodel G_{11}, which represents the relation $y_1 = f(u_1)$, $i.e.$ the belt speed y_1 as a function of the left drive motor rotation speed u_1. Measurements were implemented in the following way: the voltage of the right drive motor u_2 was kept constant and a random signal with Gaussian distribution was superposed on the working points of the input voltage u_1 in the following ranges (see Figure 3.11):

$$u_{11}(k) = 0.2 + e_s(k); \quad \text{for } k \in \langle 0; 40 \rangle \text{s}$$

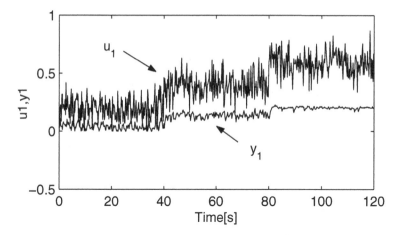

Figure 3.11. Example 3.5. Input and output signal of the identified object

$$u_{12}(k) = 0.4 + e_s(k); \quad \text{for } k \in \langle 41; 80 \rangle \text{s}$$

$$u_{13}(k) = 0.6 + e_s(k); \quad \text{for } k \in \langle 81; 120 \rangle \text{s}$$

The initial vector of the parameter estimates was chosen *ad hoc* in the form $\hat{\Theta}^T(0) = [0.1; 0.2; 0.3; 0.4]$, the sampling period $T_0 = 0.2$ s and the structure of a second-order model was chosen as a discrete transfer function

$$G(z) = \frac{b_1 z^{-1} + b_2 z^{-2}}{1 + a_1 z^{-1} + a_2 z^{-2}}$$

The parameter estimates evolution over individual intervals are shown in Figure 3.12.

Experiment results. In the first interval, $t \in \langle 0; 40 \rangle$, the discrete transfer function with parameter estimates in step $t = 40$ s has the form

$$G(z) = \frac{-0.0472 z^{-1} + 0.2000 z^{-2}}{1 - 0.4661 z^{-1} + 0.1415 z^{-2}}$$

Figure 3.13 shows the calculated step response for the first interval.

In the second interval, $t \in \langle 41; 80 \rangle$, the discrete transfer function with parameter estimates in step $t = 80$ s has the form

$$G(z) = \frac{-0.0225 z^{-1} + 0.1825 z^{-2}}{1 - 0.6150 z^{-1} + 0.0763 z^{-2}}$$

and the calculated step response is presented in Figure 3.14.

In the third interval, $t \in \langle 81; 120 \rangle$, the discrete transfer function with parameter estimates in step $t = 120$ s has the form

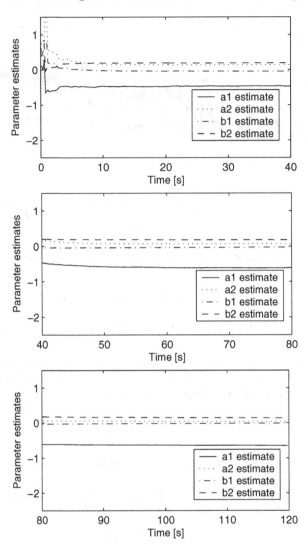

Figure 3.12. Example 3.5. Evolution of parameter estimates

$$G(z) = \frac{0.002z^{-1} + 0.1190z^{-2}}{1 - 0.6757z^{-1} - 0.0261z^{-2}}$$

and the calculated step response is shown in Figure 3.15.

From Figure 3.12 it is clear, that in the second interval the parameter estimates $\hat{a}_1, \hat{a}_2$ are changing slowly, therefore, it is possible to suppose that a controller with fixed parameters will not guarantee satisfactory control over the whole range needed by the laboratory model. From the individual discrete transfer functions and step responses in Figures 3.13–3.15 it is obvious that

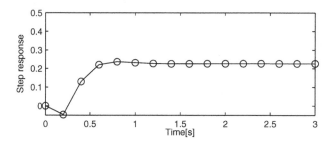

Figure 3.13. Example 3.5. Step response in the first interval

the process has nonminimum phase behaviour. Therefore the use of a suitable adaptive controller is one possibility for ensuring the control quality of the laboratory model. The design and verification of the TITO digital self-tuning controller for this laboratory model is presented in [51] and a delta modification of this controller in [52].

Example 3.6. Consider a continuous stirred tank reactor (CSTR) with first-order consecutive exothermic reactions according to the scheme $A \xrightarrow{k_1} B \xrightarrow{k_2} C$ and with a perfectly mixed cooling jacket (see Figure 3.16).

Using the usual simplifications, the model of the CSTR can be described by four nonlinear differential equations (see [53]).

$$\frac{dc_A}{dt} = -\left(\frac{Q_r}{V_r} + k_1\right) c_A + \frac{Q_r}{V_r} c_{Af}$$

$$\frac{dc_B}{dt} = -\left(\frac{Q_r}{V_r} + k_2\right) c_B + k_1 c_A + \frac{Q_r}{V_r} c_{Bf}$$

$$\frac{dT_r}{dt} = \frac{h_r}{(\rho c_p)_r} + \frac{Q_r}{V_r}(T_{rf} - T_r) + \frac{A_h U}{V_r(\rho c_p)_r}(T_c - T_r)$$

$$\frac{dT_c}{dt} = \frac{Q_c}{V_c}(T_{cf} - T_c) + \frac{A_h U}{V_c(\rho c_p)_c}(T_r - T_c)$$

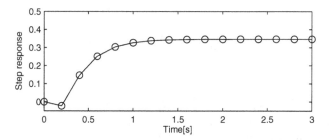

Figure 3.14. Example 3.5. Step response in the second interval

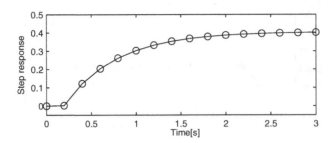

Figure 3.15. Example 3.5. Step response in the third interval

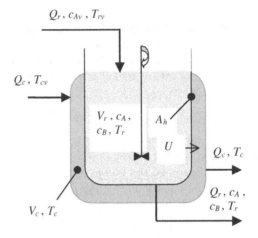

Figure 3.16. Example 3.6. Continuous stirred tank reactor

with initial conditions $c_A(0) = c_A^s$, $c_B(0) = c_B^s$, $T_r(0) = T_r^s$, and $T_c(0) = T_c^s$. Here, t is the time, c denote concentrations, T are temperatures, V stands for volumes, ρ are densities, c_p denote specific heat capacities, Q are volumetric flow rates, A_h stands for the heat exchange surface area and U is the heat transfer coefficient. The subscripts are $(.)_r$ for the reactant mixture, $(.)_c$ for the coolant, $(.)_f$ for feed (inlet) values and the superscript $(.)^s$ for steady-state values.

The reaction rates and the reaction heat are expressed as

$$k_j = k_{0j} \exp\left(\frac{-E_j}{RT_r}\right), \quad \text{for } j = 1, 2$$

$$h_r = h_1 k_1 c_A + h_2 k_2 c_B$$

where k_0 are pre-exponential factors, E denote activation energies and h stands for reaction enthalpies. The values of all parameters, feed values and steady-state values are given in Table 3.2.

For the identification purposes, the process output and input are defined as

Table 3.2. Parameters, inlet values and initial conditions

$V_r = 1.2$ m^3	$Q_r = 0.08$ m^3 min^{-1}
$V_c = 0.64$ m^3	$Q_c^s = 0.03$ m^3 min^{-1}
$\rho_r = 985$ kg m^{-3}	$c_{pr} = 4.05$ kJ kg$^{-1}K^{-1}$
$\rho_c = 998$ kg m^{-3}	$c_{pc} = 4.18$ kJ kg$^{-1}K^{-1}$
$A = 5.5$ m^2	$U = 43.5$ kJ m^{-2} min^{-1} K^{-1}
$k_{10} = 5.616 \times 10^{16}$ min^{-1}	$E_1/R = 13477$ K
$k_{20} = 1.128 \times 10^{18}$ min^{-1}	$E_2/R = 15290$ K
$h_1 = 4.8 \times 10^4$ kJ kmol^{-1}	$h_2 = 2.2.10^4$ kJ kmol^{-1}
$c_{Af} = 2.85$ kmol m^{-3}	$c_{Bf} = 0$ kmol m^{-3}
$T_f = 323$K	$T_{cf} = 293$ K
$c_A^s = 0.1649$ kmol m^{-3}	$c_B^s = 0.9435$ kmol m^{-3}
$T_r^s = 350.19$ K	$T_c^s = 330.55$ K

$$y(t) = T_r(t) - T_r^s; \qquad u(t) = 10\frac{Q_c(t) - Q_c^s}{Q_c^s}$$

These expressions enable one to obtain variables of approximately the same magnitude.

Solve the system of nonlinear differential equations in the SIMULINK® environment. Generate step changes +20%, +100% and -30% around the basic steady-state value of the volumetric flow rate $Q_c^s = 0.03$ m^3 min^{-1} with additive discrete noise. For RLSM use a second-order discrete model

$$G(z) = \frac{b_1 z^{-1} + b_2 z^{-2}}{1 + a_1 z^{-1} + a_2 z^{-2}}$$

with a sampling period $T_0 = 2.5$ minutes. Plot graphs of the evolution of model parameter estimates and prediction error for the volumetric flow rate step change +20% in time interval $\langle 0; 120 \rangle$ min.

1. Step change +20% – $Q_c^s = 1.2 \times 0.03 = 0.036$ m^3 min^{-1}.
   The z-transfer function: $G(z) = \frac{-0.0348z^{-1} - 0.0211z^{-2}}{1 - 1.6020z^{-1} + 0.6351z^{-2}}$
2. Step change +100% – $Q_c^s = 2 \times 0.03 = 0.06$ m^3 min^{-1}.
   The z-transfer function: $G(z) = \frac{-0.0345z^{-1} - 0.0179z^{-2}}{1 - 1.5546z^{-1} + 0.5859z^{-2}}$
3. Step change −30% – $Q_c^s = 0.7 \times 0.03 = 0.021$ m^3 min^{-1}.
   The z-transfer function: $G(z) = \frac{-0.0391z^{-1} - 0.0239z^{-2}}{1 - 1.5889z^{-1} + 0.6252z^{-2}}$

The evolution of parameter estimates and prediction error for the volumetric flow rate +20% in time interval $\langle 0; 120 \rangle$min are shown in Figure 3.17

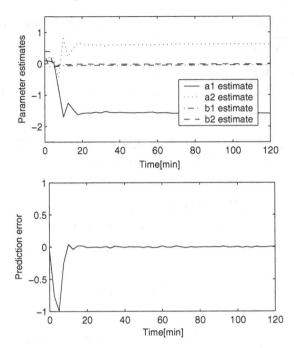

Figure 3.17. Example 3.6. Evolution of parameter estimates and prediction error – step change +20%

3.3 Summary of chapter

A recursive identification algorithm is an indispensable part of the algorithmic structure of a self-tuning controllers. The objective of this chapter was not the exact analysis of recursive identification methods but only an introduction to this area from the viewpoint of utilization of these methods in adaptive controllers. Therefore attention was paid only to the identification method most widely used in practical applications – least squares method for estimation of parameters of a simple stochastic (regressive ARX) model. For better comprehension of recursive identification methods, the theoretical discussion was complemented with several examples. One example is solved using data obtained by measurements on laboratory nonlinear TITO model CE 108 – coupled drive apparatus. For further practice in this area, an example of a analytic nonlinear model of the continuous stirred tank reactor is presented. For this analytic model, an adaptive continuous-time controller was designed and verified by simulation (see [53]).

Problems

3.1. Consider the discrete system

$$y(k) = -a_1 y(k-1) - a_2 y(k-2) + b_1 y(k-1) + b_2 y(k-2) + e_s(k)$$

where $b_1 = 0.2$, $b_2 = 0.15$, $a_2 = -0.7$ and $a_1 = 1.5$ for $0 < k < 100$; $a_1 = 0.75$ for $k \geq 100$.

Make graphs of the evolution of parameter estimates and prediction error for time interval $\langle 0; 200 \rangle$.

3.2. Consider a second-order continuous-time system

$$y''(t) + a_1 y'(t) + a_0 y(t) = b_1 u'(t) + b_0(t) u(t)$$

with $a_1 = 1.5$, $a_0 = 0.5$, $b_1 = 0.8$ and $b_0(t) = \sin t$. Use MATLAB®/SIMULINK® for simulation of this time-varying system and plot graphs of the evolution of discrete parameter estimates and prediction error for sampling period $T_0 = 0.4\,\text{s}$.

3.3. Consider a second-order continuous-time system

$$y''(t) + a_1(t) y'(t) + a_0 y(t) = b_0 u(t)$$

where $a_1(t) = \cos t$, $a_0 = 1.2$, and $b_0 = 0.5$. Use MATLAB®/SIMULINK® for simulation of this time-varying system and make graphs of the evolution of parameter estimates and prediction error for sampling period $T_0 = 0.4$ s.

3.4. Consider "interacting tanks in series process" (see Figure 3.18). The model of the process is described by two differential equations

$$F_1 \frac{dh_1}{dt} = q_0 - q_1$$

$$F_2 \frac{dh_2}{dt} = q_1 - q_2$$

The inlet q_1 depends on the difference between the liquid heights and outlet flow rate q_2 depends on the liquid height in the second tank as

$$q_1 = k_1 \sqrt{h_1 - h_2}$$

$$q_2 = k_2 \sqrt{h_2}$$

The values of all parameters, feed values and steady-state values are given in Table 3.3.

Solve the system of differential equations in the SIMULINK® environment. Identify the system by the recursive least squares method. Change the set point from the steady-state to the values $q_0 = 0.02$ m³ min⁻¹; $q_0 = 0.04$ m³ min⁻¹; $q_0 = 0.08$ m³ min⁻¹. For generation of the input signal $u(t)$, add a suitable discrete generator of random noise. Use a second-order discrete model structure for recursive identification.

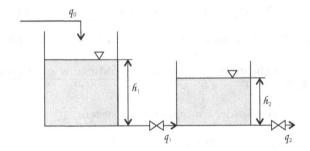

Figure 3.18. Problem 3.4 Interacting tanks in series process

where

t time [min]

h_1, h_2 heights of liquids in the first and second tanks [m]

q_0 inlet volumetric flow rate to the first tank [m^3 min^{-1}]

q_1 inlet volumetric flow rate to the second tank [m^3 min^{-1}]

q_2 outlet volumetric flow rate from the second tank [m^3 min^{-1}]

F_1, F_2 cross-section area of the tanks [m^2]

Table 3.3. Parameters, inlet values and initial conditions

$F_1 = 0.0177$ m^2	$q_0^s = 0.03$ m^3 min^{-1}
$F_2 = 0.0314$ m^2	$h_1^s = 0.36$ m
	$h_2^s = 0.25$ m
$k_1 = 0.0905$ m$^{5/2}$ min^{-1}	
$k_2 = 0.06$ m$^{5/2}$ min^{-1}	

4

Self-tuning PID Controllers

This chapter is devoted to PID digital controllers and how they may be made self-tuning.

Experience in implementing adaptive control systems indicates that users in industrial practice are mistrustful of adaptive digital controllers based on the optimal theory of automatic control, mainly because understanding the algorithms involved requires in-depth knowledge of the theory of automatic control. The common factor to all successful applications of this kind of adaptive controller has been the availability of a sufficiently qualified person in industrial practice both capable of absorbing the results of scientific research and well acquainted with the technology in use. It is for this reason that we have been able to see a trend over recent years towards research into simple adaptive controllers which can be implemented not only by theoreticians in the field of adaptive control but also by users in industrial applications.

It is clear that the vast majority of controllers (around 90%) currently used in industry are PID type controllers because, provided these are well adjusted, they show very good control results. They are also user-friendly in that they are simple, generally well known and easy to implement. Provided the parameters have been well chosen they are capable of controlling a significant portion of continuous-time technological processes.

The use of continuous-time PID controllers has a tradition going back many years. A number of adjustment approaches and methods of optimization have been developed in which users and manufacturers of control technology have wide practical experience. Understandably they wish to put this knowledge to work along with their experience with analogue techniques and apply this also to digital control systems.

The aim of this chapter is to acquaint the reader with problems of discretization of continuous-time PID type controllers. A survey of particular digital PID type algorithms which are suitable for design of their self-tuning versions is also proposed. Especially emphasized are the additional versions of PID type controllers which have a number of advantages from the point of view of their practical implementation. The proposed algorithms are partly taken from literature and partly originally designed by the authors. The chapter is accompanied by suitable examples of the design of some controllers and some simulation examples are also included to demonstrate the dynamic properties of controllers using suitable control models.

This chapter is divided into the following sections. In Section 4.1 methods of discretizing continuous-time PID controllers are described, while in Section 4.2 some modifications are discussed to improve the dynamic characteristics of controllers. Nonlinear PID controllers are the subject of Section 4.3. Section 4.4 gives a brief discussion of the issue of selecting the sampling period, and 4.5 deals with the industrial applications of PID controllers. Section 4.6 provides a survey of self-tuning controllers; 4.7 gives a detailed analysis of certain algorithms used in self-tuning PID controllers, with simulation examples of some types of controllers. Finally, in Section 4.8 simulations are performed in MATLAB® environment using SIMULINK® oriented toolbox.

4.1 PID Type Digital Controllers

The ideal, textbook version of a continuous-time PID controller is usually given in the form

$$u(t) = K_P \left[e(t) + \frac{1}{T_I} \int_0^t e(\tau)\mathrm{d}\tau + T_D \frac{\mathrm{d}e(t)}{\mathrm{d}t} \right] \tag{4.1}$$

or in the form

$$u(t) = r_0 e(t) + r_{-1} \int_0^t e(\tau)\mathrm{d}\tau + r_1 \frac{\mathrm{d}e(t)}{\mathrm{d}t} \tag{4.2}$$

where $e(t) = w(t) - y(t)$ and the conversion between (4.1) and (4.2) is

$$K_P = r_0 \qquad T_I = \frac{K_P}{r_{-1}} \qquad T_D = \frac{r_1}{K_P} \tag{4.3}$$

where $u(t)$ is the controller output, i.e. manipulated variable $y(t)$ denotes the process output, i.e. controlled variable $e(t)$ stands for the tracking error and $w(t)$ is the reference signal, i.e. set point. The parameters of the PID controller (4.1) are as follows: proportional gain K_P, integral time constant T_I, and derivative time constant T_D. The parameters for controller (4.2) are gain r_0, integral constant r_{-1}, and derivative constant r_1. Since (4.1) is the most widely used in practice, and the rules for tuning the PID controller have been defined for parameters K_P, T_I and T_D, this is the form we prefer to deal with.

Using the Laplace transform it is possible to convert Equation (4.1) into the form

$$U(s) = K_P \left[1 + \frac{1}{T_I s} + T_D s \right] E(s) \tag{4.4}$$

where s represents the Laplace transform operator. From Equation (4.4) we can determine the transfer function of the PID controller

$$G_R(s) = \frac{U(s)}{E(s)} = K_P \left[1 + \frac{1}{T_I s} + T_D s \right] \tag{4.5}$$

To obtain a digital version of a continuous-time PID controller we must discretize the integral and derivative components of Equation (4.1), see [54]. When the sampling period T_0 is small and noise from the process output signal is effectively filtered out, the simplest algorithm is obtained by replacing the derivative with a difference of the first-order (two-point, backward difference)

$$\frac{de}{dt} \approx \frac{e(k) - e(k-1)}{T_0} = \frac{\Delta e(k)}{T_0} \tag{4.6}$$

where $e(k)$ is the error value at the k-th moment of sampling, $i.e.$ at time $t = kT_0$. The easiest way of approximating the integral is by simple summing so that we approximate the continuous-time function by sampling periods T_0 of the constant function (step function, rectangle). Using the so-called forward rectangular method (FRM) yields

$$\int_0^t e(\tau) d\tau \approx T_0 \sum_{i=1}^{k} e(i-1) \tag{4.7}$$

so that the equation for a discrete PID controller has the form

$$u(k) = K_P \left\{ e(k) + \frac{T_0}{T_I} \sum_{i=1}^{k} e(i-1) + \frac{T_D}{T_0} [e(k) - e(k-1)] \right\} \tag{4.8}$$

If the continuous-time signal is discretized recursively using the step function with the help of the so-called backward rectangular method (BRM), instead of relation (4.7) we obtain relation

$$\int_0^t e(\tau) d\tau \approx T_0 \sum_{i=1}^{k} e(i) \tag{4.9}$$

and Equation (4.8) changes to form (4.10) which is most often used in the formal description of a digital PID controller

$$u(k) = K_P \left\{ e(k) + \frac{T_0}{T_I} \sum_{i=1}^{k} e(i) + \frac{T_D}{T_0} [e(k) - e(k-1)] \right\} \tag{4.10}$$

If, instead of rectangular methods (4.7), (4.9), we use the more accurate trapezoidal method (TRAP) to calculate the integral, where we replace the continuous-time signal with straight line sections, *i.e.*

$$\int_0^t e(\tau)\mathrm{d}\tau \approx T_0 \sum_{i=1}^{k} \frac{e(i) + e(i-1)}{2} \qquad (4.11)$$

then the equation for the digital PID controller will take the form

$$u(k) = K_P \left\{ e(k) + \frac{T_0}{T_I} \left[\frac{e(0) + e(k)}{2} + \sum_{i=1}^{k-1} e(i) \right] + \frac{T_D}{T_0} [e(k) - e(k-1)] \right\}$$
$$(4.12)$$

The individual methods of discretizing the integral of continuous-time error $e(t)$ at points $e(kT_0)$, where $k = 0, 1, 2, \ldots$ are illustrated in Figure 4.1.

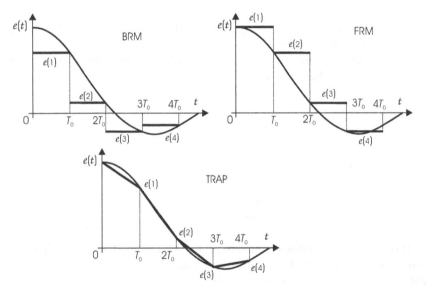

Figure 4.1. Methods of discretizing the integral component

As far as changes in error are concerned, where sampling is sufficiently fast, there is no significant difference between integral approximations and consequently between relations (4.8), (4.10) and (4.12), therefore (4.10) is the one most often used. Because the whole value of controller output $u(k)$ is calculated, usually in terms of the drive position, these algorithms are also known as *absolute* or *position algorithms* for a PID controller. Equations (4.8), (4.10) and (4.12) are called nonrecurrent algorithms, where all previous error values $e(k - i), i = 1, 2, \ldots, k$ have to be known, to calculate the integral and with it the controller action. In real industrial applications this is impractical largely because it would be necessary to keep all previous error values in the memory of the control computer. In the form they are given, Equations (4.8), (4.10) and (4.12) are also unsuitable with regards to changes in the controller parameters – a change in T_I or K_P leads to an instant change of the entire value of the integral component resulting in an overload of the whole calculation of the error, which is not acceptable. Recurrent algorithms are, therefore, more suitable for practical use. It is either necessary to calculate recurrently integral (4.9) – see algorithms (4.27) and (4.28) below – or controller output value $u(k)$ from a previously recorded value $u(k-1)$ plus correction increment $\Delta u(k)$. Alternatively, for a PID controller with digital output, just the increment (change) $\Delta u(k)$ may be calculated. Algorithms which calculate increment (change) $\Delta u(k)$ are referred to as *incremental* or *velocity algorithms*. By subtracting Equation (4.10), which we obtained from the backward rectangular method, for steps k and $k-1$, we obtain the recurrent relation

$$u(k) = \Delta u(k) + u(k-1) \tag{4.13a}$$

$$\Delta u(k) = K_P \left\{ e(k) - e(k-1) + \frac{T_0}{T_I} e(k) \right. \tag{4.13b}$$
$$\left. + \frac{T_D}{T_0} [e(k) - 2e(k-1) + e(k-2)] \right\}$$

and in general form

$$u(k) = q_0 e(k) + q_1 e(k-1) + q_2 e(k-2) + u(k-1) \tag{4.14}$$

Controller parameters q_0, q_1 and q_2 in Equation (4.14) are given in Table 4.1.

Using (4.14) it is possible to calculate the step response of PI and PID type digital controllers (Figures 4.2 and 4.3). For the step response of a digital PID controller to approach that of a continuous-time PID controller the following must be valid (for positive controller gain $q_0 > 0$):

- the second sample of controller output $u(1) < u(0)$;
- constant positive growth of the step response (from $k = 2$);
- straight line linear growth must intersect positively with ordinate axis $u(k)$ (for a continuous-time PID controller this value is $K_P = r_0$).

This corresponds to limits on values q_0, q_1 and q_2

$$q_0 > 0 \qquad q_1 < -q_2 \qquad -(q_0 + q_1) < q_2 < q_0 \tag{4.15}$$

Value q_0 determines the size of first action $u(0)$ for the step-change of the reference signal w and the initial zero steady-state.

Recurrent relation (4.14) obtained from (4.10) (BRM) can also be written in the form

$$u(k) = K_P \left\{ e(k) - e(k-1) + \frac{T_0}{T_I} e(k) \right.$$
$$\left. + \frac{T_D}{T_0} [e(k) - 2e(k-1) + e(k-2)] \right\} + u(k-1) \tag{4.16}$$

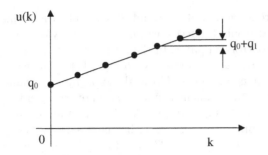

Figure 4.2. Step response of a digital PI controller

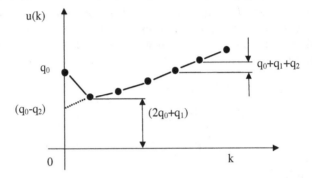

Figure 4.3. Step response of a digital PID controller

which is essentially form (4.13a). The advantage of form (4.14) is the simple structure of the algorithm but it has the disadvantage of having a less transparent link between the individual components. Comparing (4.14) and (4.16) yields

$$K_P = q_0 - q_2 \qquad \frac{T_D}{T_0} = \frac{q_2}{K_P} \qquad \frac{T_0}{T_I} = \frac{q_0 + q_1 + q_2}{K_P} \qquad (4.17)$$

To continue, we give the incremental algorithm deduced from Equation (4.8) (FRM)

$$
\begin{aligned}
u(k) = K_P \Big\{ &e(k) - e(k-1) + \frac{T_0}{T_I} e(k-1) \\
&+ \frac{T_D}{T_0} [e(k) - 2e(k-1) + e(k-2)] \Big\} + u(k-1)
\end{aligned}
\qquad (4.18)
$$

The incremental algorithm derived from Equation (4.12) (TRAP) has the form

$$
\begin{aligned}
u(k) = K_P \Big\{ &e(k) - e(k-1) + \frac{T_0}{2T_I} [e(k) + e(k-1)] \\
&+ \frac{T_D}{T_0} [e(k) - 2e(k-1) + e(k-2)] \Big\} + u(k-1)
\end{aligned}
\qquad (4.19)
$$

and the individual equation parameters for form (4.14) are once again shown in Table 4.1.

Table 4.1. Parameters of a digital incremental PID controller

Controller Parameters	FRM	BRM	TRAP
q_0	$K_P \left(1 + \frac{T_D}{T_0}\right)$	$K_P \left(1 + \frac{T_0}{T_I} + \frac{T_D}{T_0}\right)$	$K_P \left(1 + \frac{T_0}{2T_I} + \frac{T_D}{T_0}\right)$
q_1	$-K_P \left(1 - \frac{T_0}{T_I} + \frac{2T_D}{T_0}\right)$	$-K_P \left(1 + \frac{2T_D}{T_0}\right)$	$-K_P \left(1 - \frac{T_0}{2T_I} + \frac{2T_D}{T_0}\right)$
q_2	$K_P \frac{T_D}{T_0}$	$K_P \frac{T_D}{T_0}$	$K_P \frac{T_D}{T_0}$

From Table 4.1 it follows that incremental controller parameters q_0, q_1 and q_2 expressed in the form of Equation (4.14) are functions of proportional gain K_P, time constant integral T_I and derivative T_D, sampling period T_0 and the method of discretization, $i.e.$ the functional relations

$$q_i = f(K_P, T_I, T_D, T_0); \quad \text{for } i = 0, 1, 2 \tag{4.20}$$

are valid.

Some equations for digital PID controllers can also be derived using transformation formulas which allow us to transfer continuous-time description from the Laplace transform to discrete transfer functions. These transformation formulas take the form

$$s = \frac{1 - z^{-1}}{z^{-1} T_0} \qquad s = \frac{1 - z^{-1}}{T_0} \qquad s = \frac{2}{T_0} \frac{1 - z^{-1}}{1 + z^{-1}} \tag{4.21}$$

where z^{-1} is the backward time-shift operator, $i.e.$ $x(k - 1) = z^{-1} x(k)$. Using the second formula in (4.21) it is possible to obtain controller (4.16) from transfer function (4.5). The use of the third formula of (4.21) to obtain further modifications of a digital PID controller is explained in Section 4.2.1.

Niederlinski [55] has discussed the properties of positional (absolute) and incremental controllers with regard to their practical implementation, but the conclusions reached apply mainly to the automation and computer technology available at the time of writing.

The approximation of the derivation term can also be performed by replacing the derivation by a four-point mean difference using the relation

$$\Delta e(k) = \frac{1}{6}[e(k) + 3e(k - 1) - 3e(k - 2) - e(k - 3)] \tag{4.22}$$

The position algorithm obtained using the BRM to discretize the integral component then has the form

$$u(k) = K_P \left\{ e(k) + \frac{T_0}{T_I} \sum_{i=1}^{k} e(i) \right.$$
$$\left. + \frac{T_D}{6T_0} [e(k) + 3e(k-1) - 3e(k-2) - e(k-3)] \right\} \quad (4.23)$$

and the incremental algorithm takes either form

$$u(k) = K_P \left\{ e(k) - e(k-1) + \frac{T_0}{T_I} e(k) \right.$$
$$\left. + \frac{T_D}{6T_0} [e(k) + 2e(k-1) - 6e(k-2) + 2e(k-3) + e(k-4)] \right\} \quad (4.24)$$
$$+ u(k-1)$$

or

$$u(k) = q_0 e(k) + q_1 e(k-1) + q_2 e(k-2) + q_3 e(k-3) + q_4 e(k-4) + u(k-1) \quad (4.25)$$

where

$$q_0 = K_P \left(1 + \frac{T_0}{T_I} + \frac{T_D}{6T_0} \right) \qquad q_1 = -K_P \left(1 - \frac{T_D}{3T_0} \right)$$
$$q_2 = -K_P \frac{T_D}{T_0} \qquad q_3 = K_P \frac{T_D}{3T_0} \qquad q_4 = K_P \frac{T_D}{6T_0} \quad (4.26)$$

Instead of position algorithm (4.10) it is possible to use a component form of the control algorithm, where the controller output is determined by the sum of the individual components and only the last value of the integral component is retained in the memory. In this case the algorithm takes the form

$$u(k) = u_P(k) + u_I(k) + u_D(k) \quad (4.27)$$

where

$$u_P(k) = K_P e(k)$$
$$u_I(k) = K_P \frac{T_0}{T_I} \sum_{i=1}^{k} e(i) = u_I(k-1) + K_P \frac{T_0}{T_I} e(k) \quad (4.28)$$
$$u_D(k) = K_P \frac{T_D}{T_0} [e(k) - e(k-1)]$$

are the controller's proportional, integral and derivative components.

The advantage of the component form is its transparency when setting the gain of the individual controller components. However, it is necessary to limit integral component u_I in the algorithm – see Section 4.5.2.

The definition of initial values $u_I(0)$ for (4.28) and $u(0)$ for algorithms (4.16), (4.18), (4.19) and (4.27) and the differences in the behaviour between these formally congruent algorithms are treated in Sections 4.5.1 and 4.5.2. Also compared are various PID algorithms from the point of view of the impact on controller output $u(k)$ if parameters K_P, T_I and T_D change.

From the point of view of nearly the same dynamics, the approximation of
a continuous-time PID controller is only suitable where the sampling period
T_0 is short compared with system dynamics. For a greater value of T_0, the
simple transfer of parameters K_P, T_I and T_D from a continuous-time to a
digital controller, even in conjunction with period T_0, is unacceptable and
all parameters must be set for the given sampling period T_0. If we take into
account that for the same sampling period T_0 roughly the same energy $\sum u$
is put into the system in order to regulate it, we can permit a significantly
larger control output at higher levels of sampling, and therefore greater gain
K_P, than we can for low levels of sampling. Generally, it is possible to state
that with an increase in sampling period T_0, gain K_P nonlinearly falls, term
$K_P T_0/T_I$ slightly increases, while term $K_P T_D/T_0$ grows smaller.

4.2 Modifying Digital PID Controllers

While Equation (4.1) represents an idealization of a real PID-like controller
behaviour, computation of the control law for digital PID-like controllers proceeds according to a corresponding difference equation. This fact causes some
problems with practical implementation of these controllers, since in contradis-
tinction to continuous-time controllers, here natural suppression of large and
rapid changes of control errors and controllers outputs (manipulated) variables is not enabled. In a particular realization of the continuous-time PID-
like function a delay appears, which is omissible owing to the dynamics of the
controlled process and it is not necessary to take it into account. But this delay
operates both like a natural noise filter (namely its higher frequency elements)
and delaying factor during step changes of reference signals. Ideal step changes
of signals for continuous-time are unrealizable anyway. That is why the risk of
rapid changes of manipulated variables connected with dynamical straining of
an actuator is not so high in continuous-time control loops as in discrete control loops where discrete controllers get and accept step changes of reference
signals as changes of variable in a program. The variable acquires a new value
of the controller output according to Equation (4.14). Large changes of the
manipulated variable in a discrete controller follow large changes of control
error. Then the contribution of the proportional and derivative parts to the
general function of a PID-like controller is significant. Since the controlled
variable is influenced by noise, samples of the controlled variable are affected
by random faults. The changes of control error cause not only changes of controller output variable but also changes given by the faulty process output
variable.

The negative effect of the large and rapid changes of control error can be
suppressed by incorporation of specific technical or software precautions ahead
of the algorithm for computation of the discrete PID function (preprocessing of
$e(k)$). Another possibility is modification of the default discrete PID algorithm

so that some of the filtering or delaying precautions are incorporated directly in the algorithm.

4.2.1 Filtering the Derivative Component

There is often interference in the measured value of the process output $y(t)$ from relatively high frequency noise. If a derivative component is used in the controller, the derivation of the signal affected by noise, which has, in addition, been approximated by a simple differential of the first-order, may cause inappropriate and unsuitably large changes in the controller output. The derivative component is limited, therefore, either by using a limiter or, more often, a first or second-order filter which decrease gain at higher frequencies. When using a first-order filter (single capacity filter) with time constant T_f the derivative component takes the form

$$D(s) = K_P \frac{T_D s}{T_f s + 1} E(s) \qquad T_f = \frac{T_D}{\alpha}; \qquad \alpha \in \langle 3; 20 \rangle \qquad (4.29)$$

Usually $\alpha = 10$ is selected which means that the D-component filter has a time constant ten times smaller than the derivative time constant. Discretization (4.29) using backward rectangular integration (the second transformation formula of (4.21)) gives the relation

$$d(k) = \frac{T_D d(k-1) + K_P T_D \alpha [e(k) - e(k-1)]}{T_D + \alpha T_0} \qquad (4.30)$$

Using the Tustin transformation (the third transformation formula of (4.21)) yields

$$d(k) = \frac{(2T_D - \alpha T_0)d(k-1) + 2K_P T_D \alpha [e(k) - e(k-1)]}{2T_D + \alpha T_0} \qquad (4.31)$$

Both approximations (4.30) and (4.31) have the form

$$d(k) = ad(k-1) + b[e(k) - e(k-1)] \qquad (4.32)$$

but with differing a, b coefficients. Approximations (4.30) and (4.31) are stable for all $T_D > 0$ situations. However in (4.31) the coefficient is $a < 0$ when $T_D < \alpha T_0/2$ which can cause unwanted oscillation during calculation especially if $T_D \ll T_0$. Only (4.30) provides good results for all values of T_D.

It is possible to use relation (4.30) or (4.31) in previous formulas (for example in (4.27)) in which the existing expression for the derivative is replaced.

Another complex method of obtaining a digital PID controller with D-component filtration will be presented. The transfer function of the continuous-time version of this type of PID controller has the form

$$G_R(s) = \frac{U(s)}{E(s)} = K_P \left(1 + \frac{1}{T_I s} + \frac{T_D s}{T_f s + 1} \right) \qquad (4.33a)$$

where T_f is the time constant of the derivative component filter. We use the third transformation formula of (4.21) (the Tustin approximation) to discretize (4.33a). The discrete form of controller transfer function (4.33a) is given by

$$G_R(z) = \frac{U(z)}{E(z)} = \frac{Q(z^{-1})}{P(z^{-1})} \tag{4.33b}$$

where

$$Q(z^{-1}) = q_0 + q_1 z^{-1} + q_2 z^{-2} \qquad P(z^{-1}) = 1 + p_1 z^{-1} + p_2 z^{-2} \tag{4.34}$$

The equation for the controller with filtration of the D-component then takes the form

$$u(k) = p_1 u(k-1) + p_2 u(k-2) + q_0 e(k) + q_1 e(k-1) + q_2 e(k-2) \tag{4.35}$$

where the parameters of controller (4.35) are given by

$$
\begin{aligned}
p_1 &= \frac{\frac{-4T_f}{T_0}}{\frac{2T_f}{T_0}+1} \qquad p_2 = \frac{\frac{2T_f}{T_0}-1}{\frac{2T_f}{T_0}+1} \\
q_0 &= \frac{K_P + 2K_P\frac{T_f+T_D}{T_0} + \frac{K_P T_0}{2T_I}\left(\frac{2T_f}{T_0}+1\right)}{\frac{2T_f}{T_0}+1} \\
q_1 &= \frac{\frac{K_P T_0}{2T_I} - 4K_P\frac{T_f+T_D}{T_0}}{\frac{2T_f}{T_0}+1} \\
q_2 &= \frac{\frac{T_f}{T_0}\left(2K_P - \frac{K_P T_0}{T_I}\right) + 2\frac{K_P T_D}{T_0} + \frac{K_P T_0}{2T_I} - K_P}{\frac{2T_f}{T_0}+1}
\end{aligned}
\tag{4.36}
$$

4.2.2 Supression of Large Changes in the Controller Output

Instead of using error $e(k)$ in the derivative component, we can use process output $y(k)$ to decrease the larger changes in the controller output resulting from set point changes. In this case algorithm (4.16) (and similarly (4.18) and (4.19)) has the form

$$
\begin{aligned}
u(k) = K_P &\left\{ e(k) - e(k-1) + \frac{T_0}{T_I}e(k) \right. \\
&\left. + \frac{T_D}{T_0}[2y(k-1) - y(k) - y(k-2)] \right\} + u(k-1)
\end{aligned}
\tag{4.37}
$$

In this way we can achieve a significant decrease in the controller output at the moment of a set point change and then a decrease in the limitation on the controller output and the movements of the final control element into an area of nonlinearity. Usually, the rise time of the process output is slowed down and overshoot is significantly decreased while the settling time remains roughly the same. The adjustment of the parameters for controller (4.37) to changes in control and disturbance differs little from the adjustment of a controller using error in the derivation.

Changes in controller output amplitude decrease further if the reference signal $w(k)$ is substantial only in the integral component.

$$
\begin{aligned}
u(k) = K_P \bigg\{ &-y(k) + y(k-1) + \frac{T_0}{T_I}[w(k) - y(k)] \\
&+ \frac{T_D}{T_0}[2y(k-1) - y(k) - y(k-2)] \bigg\} + u(k-1)
\end{aligned}
\tag{4.38}
$$

which is the well-known relation given by Takahashi [56].

Changing the process output to the reference signal is then mainly regulated by the integral component. This can, however, be a fairly slow process. To decrease larger changes in the controller output (as a result of the reference change) it is therefore useful to modify values $w(k)$ with a single-capacity filter, or a change limiter, or to employ term $\beta w(k) - y(k)$ instead of term $w(k) - y(k)$, in the proportional component, where weighting factor β is determined by the dynamics of the system and is chosen from the interval $0 < \beta < 1$. In [57] and [58] it was proved that a good characteristic of the process dynamics is the so-called normalized gain κ, which is defined as the product of the gain of the controlled process K_S and critical proportional gain K_{Pu}, where the control loop is on the point of stability

$$
\kappa = K_S K_{Pu}
\tag{4.39}
$$

Then it is possible to change PID controller parameters K_P, T_I and T_D in relation to the size of normalized gain κ. In order to reduce the maximum overshoot of the process output, the reference signal w in the proportional component (4.39) can be weighted using the factor β so that a change of the normalized gain κ is achieved. The proportional part of controller (4.27) then takes the form

$$
u_P(k) = K_P[\beta w(k) - y(k)]
\tag{4.40}
$$

As a result, the following continuous-time controller algorithm was developed which, as well as using weighting factor β, also makes use of single-capacity filter (4.29) to filter the derivative component.

$$
u(t) = K_P \left[\beta w(t) - y(t) + \frac{1}{T_I} \int_0^t e(\tau)d\tau - T_D \frac{dy_f}{dt} \right]
\tag{4.41}
$$

where $y_f(t)$ is the process output filtered by first-order transfer function

$$
\frac{Y_f(s)}{Y(s)} = \frac{1}{1 + s\frac{T_D}{\alpha}}
\tag{4.42}
$$

where the filter constant α is selected from the interval according to (4.29). The equation for a digital incremental PID controller, taken from equations (4.41) and (4.42) after replacing the derivation of the first differential and the approximation of the integral with the trapezoidal method, has the form given (see [59])

$$u(k) = u_{PI}(k) + u_D(k) \qquad (4.43)$$

where

$$u_{PI}(k) = K_P[y(k-1) - y(k)] + \frac{K_P T_0}{2T_I}[e(k) + e(k-1)] \qquad (4.44)$$
$$+ \beta K_P[w(k) - w(k-1)] + u_{PI}(k-1)$$

$$u_D(k) = K_P \frac{T_D \alpha}{T_D + T_0 \alpha}[y(k-1) - y(k)] + \frac{T_D}{T_D + T_0 \alpha} u_D(k-1) \quad (4.45)$$

4.3 Nonlinear PID Controllers

Real control loops often contain various nonlinearities in the final control elements as well as in the controlled system. Further nonlinear elements are sometimes added to improve the dynamic behaviour and/or decrease the influence of current nonlinear elements. Therefore, we will give some reasons for introducing nonlinear PID controllers and nonlinearities into the control loops:

- The nonlinearity of the controlled system (gain and/or time constants – a controller with nonlinear gain is useful for controlling the level of liquid in a spherical tank, for example, where the set point is not constant but subject to large changes; the same changes of flow cause different changes in level depending on the current level).
- The nonlinear characteristics of the final control element, usually a valve (for example flow or pressure depends on the valve position) – the nonlinear transformation of the controller output is used (inverse nonlinearity to the nonlinearity of the valve).
- Measurement noise in the controlled variable.
- Nonlinearities in the actuator, mainly dead zone, hysteresis and saturation.

- Attempts to improve the dynamic behaviour of the control loop affected by great and small disturbances.

The effect of nonlinear elements depends also on their position in the control loop. Static nonlinear elements like saturation or insensitivity in feedback have in fact an "inverse" characteristic then in the direct path. Hysteresis (dead band) worsens control responses and results in oscillating responses with amplitude dependent on hysteresis band. Controlled variables (process outputs) usually have higher overshoots after set point changes compared with

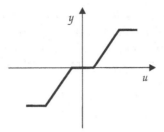

Figure 4.4. Dead zone with limitations

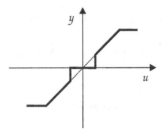

Figure 4.5. Static nonlinearity

the linear case. The effect of process output noise can be diminished by introducing insensitivity, *i.e.* a nonlinear element, dead zone, into the controller output or into the controller input (see Figure 4.4). Dead zone as well as insensitivity caused by signal quantization decrease the effect of small-amplitude noise and can damp down control responses. Similarly, the nonlinearity illustrated in Figure 4.5 gives a range of controller output over which the actuator is unable to react and may significantly damp the actuator. However such insensitivity can lead to (permanent) nonzero control error and also to (permanent) oscillations if a PI controller is used – the integral component varies up to band level and then control action is applied. The limitation of controller output included in a controller algorithm or limitation of the control error cause slower control responses in general, but not always (see Figure 4.14 in Section 4.5.2). Limitation outside of a controller and not considered in the controller algorithm (existing in the actuator, final control element *etc.*) results in complicated control problems – increase of the integral term (*"wind-up"* effect), overshoots of process output *etc.*

The use of a nonlinear controller with variable gain has a similar effect to introducing insensitivity into the controller output and, for the sake of clarity, we give it here in its continuous-time version

$$u(t) = K_P f(e) \left[e(t) + \frac{1}{T_I} \int_0^t e(\tau) \mathrm{d}\tau + T_D \frac{\mathrm{d}e(t)}{\mathrm{d}t} \right] \qquad (4.46)$$

where

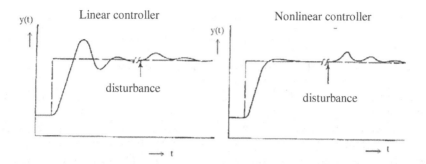

Figure 4.6. Linear and nonlinear PID controller with variable gain according to Equation (4.46)

$$f(e) = K_0 + (1 - K_0)|e(t)| \tag{4.47}$$

where K_0 is selected at interval $\langle 0; 1 \rangle$. When $K_0 = 1$ the controller is linear; when K_0 approaches zero the proportional term will be proportional to the square of the error and the controller will be insensitive to small errors. This may cause offset of control error. The controller shown may be useful in systems where gain changes in indirect proportion to amplitude, or where the controlled variable is affected by noise which causes problems in using the derivative term. In comparison with linear controllers, the nonlinear controller improves the response of controlled variable $y(t)$ to set point changes (Figure 4.6), but makes the response $y(t)$ worse when significantly affected by disturbance. This is because the effects of the disturbance result in slow drift of the controlled variable away from the set point, where a nonlinear controller has a smaller gain than a linear one.

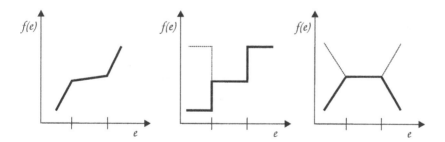

Figure 4.7. Various forms of nonlinear gain $f(e)$

Function $f(e)$ can also be defined by sections as shown in Figure 4.7. A common variation on the nonlinear controller is the controller with variable gain, defined by function $f(e)$ as follows:

$$f(e) = 1 \quad \text{for } e > e_u$$
$$f(e) = 1 \quad \text{for } e < e_l \tag{4.48}$$
$$f(e) = K_0 \quad \text{for } e \in \langle e_l; e_u \rangle, \ K_0 \geq 0$$

The output value for $K_0 = 0$ at points e_l, e_u will not change and this may significantly damp the actuator.

Another more generalized variant – see Figure 4.8 – is a controller with gain which changes linearly according to signal $v(t)$ outside fixed limits v_l, v_u; inside these limits gain is constant and equal to one. The total controller gain is $K_p = K f(v)$. Then the relation for gain can be written in the form

$$f(v) = 1 + (K_1 - 1)(v_l - v)/10 \quad \text{for } v < v_l, \tag{4.49a}$$
$$f(v) = 1 + (K_2 - 1)(v - v_u)/10 \quad \text{for } v > v_u, \tag{4.49b}$$
$$f(v) = 1 \quad \text{for } v \in \langle v_l; v_u \rangle \tag{4.49c}$$

where K_1 and K_2 are the values of $f(v)$ at points 10% from the appropriate limit v_l (from the left) or v_u (from the right). Value $v(t)$ lies in the range $\langle 0; 100 \rangle [\%]$, and the total controller gain for $v \in \langle v_l; v_u \rangle$ is $K_p = K$.

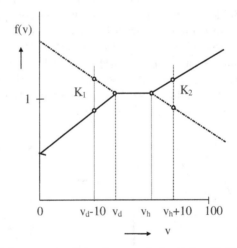

Figure 4.8. Gain characteristic $f(v)$ – (4.49)

The following can be considered as signal $v(t)$:

- input into the system;
- process output (controlled variable);
- control error;
- controller output;
- external variables;

- in some cases, a logic signal.

In a similar way it is possible to introduce a nonlinear change into the integral constant.

The dynamic systems with variable system gain defined by any function of $f(v)$ are often controlled by nonlinear PID controllers with "gain scheduling"; the controller gain approximating inverse function $f^{-1}(v)$ is defined by several segments.

Note

The controllers with variable gain outlined here are often, but inaccurately, referred to as controllers with adaptive gain in some alternative publications.

Every gain change (switch) in general results in a step in the controller output $u(k)$. The size of the step strictly depends on the applied version of PID algorithm – compare the proportional term in the incremental algorithm (4.16) based on difference of $e(k), e(k-1)$ and in the component-form algorithm (4.27) and (4.28) where only value $e(k)$ is used.

It is occasionally problematic to set the parameters of gain functions (4.47)–(4.49 for the nonlinear controllers given here when they are used to control almost linear systems. Note that the introduction of nonlinear parameters leads to controllers with variable structure. Such controllers were popular and in use in previous years. From this point of view the controllers with gain functions (4.47)–(4.49) *etc.* are only a poor approximation of controllers with variable structure. Šindelář [60, 61], for example, has performed an analysis on this as well as the conditions for switching over on the basis of control error and its derivation.

The control responses may be improved if not function $f(e)$ but a function of the weighted sum of control error and its derivation is used in (4.48).

4.4 Choice of Sampling Period

The calculation time for digital controllers is usually between a hundredth of a second and one second. Although the sampling period is one of the controller parameters, the shortest period is often used so that, effectively, these digital PID controllers may be considered to be continuous-time (more precisely: when control is applied to a system for which the controller period is $T_0 \ll T_a$, where T_a, denotes the dominant time constant of the system). Nevertheless it is wise to be aware of the influence the choice of period T_0 has on control and of the difference compared to a continuous-time controller.

In general, shortening sampling period T_0 improves the quality of control, the ability to react to disturbances and causes a discrete digital controller to resemble a continuous-time one. However, it also increases the demands on the actuator (step changes in controller output are generated with period T_0) and there is usually an increase in the system's energy consumption, judged according to the quadratic criteria

$$J_u = \int u^2(t)\mathrm{dt} \tag{4.50}$$

Lengthening sampling period T_0 often slows the control process slightly but decreases change in the manipulated variable and the value of criteria (4.50) significantly. In particular, lengthening sampling period T_0 should lower gain value K_P – see reference in Section 4.1. However, shortening the sampling period can cause deterioration of control to the point of instability even for small controller gain. This is due to the fact that discretization of the controlled system causes a shift from a so-called minimal phase system to a nonminimum phase system which has unstable zeros (that is the roots of the discrete transfer function numerator of the controlled system are outside the unit circle in the z complex plane). This fact is demonstrated by, for example, Roberts [62] who optimized the parameters of a digital PID controller using ITAE criteria

$$J = \int\limits_0^\infty t|e(t)|\mathrm{dt} \tag{4.51}$$

for a system approximated by the transfer

$$G(s) = \frac{K_S}{T_1 s + 1}\mathrm{e}^{-\tau_d s} \tag{4.52}$$

The dependence of the gain $K = K_S K_P$ (K_S is the system gain) and of the value of criterion J on the sampling period (ratio T_0/T_1) and on the size of the time delay (ratio τ_d/T_1) is illustrated in Figure 4.9 (from Roberts [62]). Only the PID controller gain K_P for integral time constant $T_I = 5T_0$ and derivative time constant $T_D = 0.8T_0$ for step changes of the reference signal are optimized. When choosing the sampling period it is recommended to use: $\omega_u T_0 < \pi/4$ where ω_u is the critical frequency at which an open-loop frequency characteristic first crosses a negative semi-axis. Figure 4.9 shows that the controller is very sensitive, especially at high sampling rate, *i.e.* when the ratio T_0/T_1 is low and the time delay small (ratio $\tau_d/T_1 < 0.5$). It is necessary to remember to include "artificial" time delay caused by the controller calculation time in the overall time delay.

The choice of the sampling period is mainly affected by:

1. The required standard of control, which is given by the demands on the controlled variable response, the control error band, and the response and changes in the manipulated variable.
2. The dynamics of the controlled system, characterized, for example, by the value of the approximating time constant and time delay according to (4.52).

3. The disturbance frequency spectrum. It follows from the behaviour of the control dynamic factor $S(\omega)$ (Figure 4.10), defined as the ratio between the amplitudes of closed and open-loop frequency transfer functions (that is the ratio between a controlled and an uncontrolled loop) that the effect of low-frequency disturbance on control is suppressed, high-frequency disturbance passes through almost without control and medium frequency is magnified (see Figure 4.10). If control is to suppress disturbances up to the frequency level ω_{max}, then the Shannon–Kotelnik theorem must be used to select the sampling period T_0

$$T_0 \leq \frac{\pi}{\omega_{max}} \tag{4.53}$$

4. The calculation time demands and by the capabilities of the computer employed, the number of control loops, etc.
5. Demands on operator intervention, thus limiting the maximum sampling period T_0.
6. The properties of the actuator – in drives this means dead zone, response time and permitted number of switches per hour (limits the minimum sampling period T_0); similarly for contactors, circuit breakers, relays, etc.

A summary of the rules for choosing the sampling period has been taken from Isermann [54] and is shown in Table 4.2.

Table 4.2. Summary of the rules for choosing the sampling period

Criteria to determine the sampling period T_0	Determining relation	Notes
	$T_0 = \left(\frac{1}{8} \div \frac{1}{16}\right)\frac{1}{f}$	Systems with dominant
	$T_0 = \left(\frac{1}{4} \div \frac{1}{8}\right)\tau_d$	time delay (transport lag)
Takes 15% longer settling time than a continuous-time control circuit with PID controller	$T_0 = (0.35 \div 1.2)T_u$	$0.1 \leq T_u/T_0 \leq 1.0$
Compensation of disturbances up to ω_{max} as in continuous-time control circuit	$T_0 = \frac{\pi}{\omega_{max}}$	$|G_P(j\omega_{max})| = 0.01 \div 0.1$
Simulation results	$T_0 = \left(\frac{1}{6} \div \frac{1}{15}\right)T_{95}$	

Comments:
f – natural frequency of the closed loop [Hz],
τ_d – time delay,
T_u – apparent dead time,
T_{95} – 95% of the step-response settling time,
G_P – system transfer function

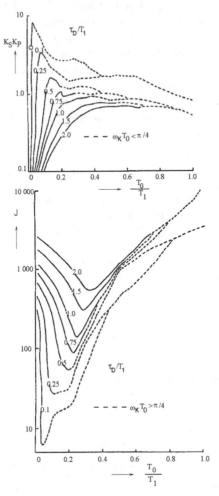

Figure 4.9. Dependence of optimum gain $K = K_S K_P$ and minimum criteria ITAE (4.51) on the sampling period of a digital PID controller (from Roberts [62])

Considering the speed of presentday microprocessors, the most significant limit placed on T_0 is, in fact, point 6, but again we should remember that from the point of view of "dulcification" a controlled and manipulated variable, longer sampling may be better than shorter sampling.

The choice of sampling period is also affected by the precision of the A/D and D/A converters used, the length of the computer words (see Section 4.5.3); and the size of controller time constants T_I and T_D also play an important role. If the ratio T_D/T_0 is too large, the controller will react more to noise causing pulsing (steps in the controller output), and thus oscillation in the process output. Shinskey [63] recommends the ratio

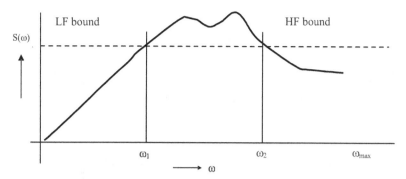

Figure 4.10. Behaviour of the control dynamic factor

$$T_D/T_0 = 0.17\omega_n/T_0 \in \langle 10; 20 \rangle$$

which leads to $T_0 \in \langle 0.008\omega_n; 0.017\omega_n \rangle$ where ω_n is the natural frequency of the system. If T_0/T_I is too small, a permanent control error will occur as a result of the limited precision of the converter, computer and rounding errors (see Section 4.5.3).

4.5 PID Controllers for Operational Use

The continuous-time PID controllers given in the available reference works are almost exclusively in form (4.1), (4.2), or (4.4) and, after discretization, in form (4.8), (4.10), (4.12) or in form (4.16), (4.18), (4.19) and (4.27). These relations are often mechanically programmed which usually results in the controller not operating well in practice. The reasons behind the malfunction are the unproved but tacit assumption (though unsatisfied) of the linearity of the control system, limitless controller output, zero initial steady state, zero initial set point, error-free calculation, *etc.*

The real simple control loop shown in Figure 4.11 includes actuator S_S, with nonlinearities such as limits on position and speed, dead zone *etc.*, actuator controller R_S (final amplifier – P controller), and own controlled system S_H (with nonlinearity of the control device) controlled by PID controller R_H. The R_H controller output, that is the calculated manipulated variable $u(k)$, is the set point for the inner control loop (servo-loop), *i.e.*

$$w^S = u^H = u \tag{4.54}$$

The measured manipulated variable u^m of controller R_H is the input for S_H, which is simultaneously the output of the inner loop, *i.e.*

$$u^m = u^{mH} = y^S \tag{4.55}$$

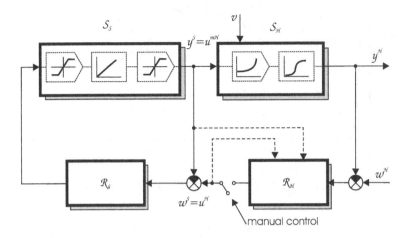

Figure 4.11. Simple real control loop

4.5.1 Initial Conditions and Controller Bumpless Connection

The digital PID controller described in component form, Equation (4.27), can be written with a separate integral component u_I in the form

$$u_I(k) = u_I(k-1) + r_{-1}e(k) \tag{4.56}$$

$$u(k) = r_0 e(k) + u_I(k) + r_1[e(k) - e(k-1)] \tag{4.57}$$

where

$$r_0 = K_P \qquad r_{-1} = \frac{K_P T_0}{T_I} \qquad r_1 = \frac{K_P T_D}{T_0} \tag{4.58}$$

or, for incremental controller (4.16), in the form

$$\Delta u_I(k) = r_{-1}e(k) \tag{4.59}$$

$$\Delta u(k) = r_0[e(k) - e(k-1)] + \Delta u_I(k) \\ + r_1[e(k) - 2e(k-1) + e(k-2)] \tag{4.60}$$

Then, the controller output is calculated using relation

$$u(k) = u(k-1) + \Delta u(k) \tag{4.61}$$

Here, for the sake of simplicity, the form with parameters r_0, r_{-1} and r_1 is used.

The initial values which are missing in (4.1) and therefore also in (4.56)–(4.61) are

$$e(-1) = e(0) \qquad y(-1) = y(0) \qquad w(-1) = w(0) \qquad u(0) \tag{4.62}$$

and particularly

$$u_I(0) = r_{-1} \int_{-\infty}^{0} e(\tau)\mathrm{d}\tau = u(0) - r_0 e(0) - r_1[e(0) - e(-1)] \tag{4.63}$$

$$u(0) \approx u^m(0)$$

where $u(0)$ indicates the initial (steady-state) value of the controller output (equal to the measured value u^m) which corresponds to the initial steady-state value of the process output $y(0)$ (time $k = 1$ is regarded as the first step in the controller calculation). Value $u(0)$ can be considered to be the reference value related to the output of controller (4.57) or (4.61).

If the controller does not contain the integral component (*i.e.* parameter $r_{-1} = 0$ and it is a P, PD or D type controller), then Equation (4.57) must contain either the term which, for uniformity of notation, will be referred to as u_I, that is

$$u_I(k) = u_I(0) = u(0) - r_0 e(0) = \text{const} \tag{4.64}$$

or term (4.63) which also includes the derivative component.

Term (4.64) acts as a bias and is usually nonzero, *i.e.* in static systems the steady-state value of the controlled variable corresponds to the steady-state nonzero value of the controller output. Term (4.64) is often referred to as "bias" or "manual reset", whereas the integral term, which changes automatically according to the control error, is known as "reset" or "automatic reset". The commonly assumed zero value in (4.62) and the absent or zero-value term (4.63) or (4.64) leads to the incorrect calculation of $u(k)$ and step $u(1)$.

To ensure bumpless transition when switching from manual to automatic control, controller output R_H – see Figure 4.11 – should track the value x_k, otherwise defined as

$$x(k) = \begin{cases} w^S(k) & \text{... slave loop set point (manual control signal)} \\ u^m(k) & \text{... measured manipulated variable} \end{cases} \tag{4.65}$$

Note

Set point w^S for manual control is often given directly from the controller under manual control; controller output is then $u = w^S$. In the loop shown in Figure 4.11, variable u^m equals the output from the final control element or actuator position y^S.

Incremental controller (4.59), (4.60) with output $\Delta u(k)$ is bumpless because it only uses the difference $\Delta u(k)$ from the current status. This is particularly useful when controlling the actuator directly from the digital controller outputs.

For incremental controller (4.59)–(4.61) with output $u(k)$

$$u(k-1) = x(k) \tag{4.66}$$

Output from position controller (4.56), (4.57) is set according to (4.65). Unlike incremental controller (4.59)–(4.61), the initial value of integral term $u_I(k-1)$ must still be defined for (4.56) and bias (4.64) for any controller without integral term

$$u_I(k-1) = x(k) - r_0 e(k-1) \qquad (4.67)$$

which means that, on connection, $u(k)$ changes mostly by the derivative term (for nonzero value r_1).

Provided the control error is nonzero and roughly constant when switching from manual to automatic control, position controller (4.57) with its integral term will alter the output using primarily the integral term (so as to achieve zero control error), which means sometimes quite a slow process (due to the size of parameter r_{-1} and control error). In this case a controller without an integral term (P, PD) leaves the output (nearly) unchanged.

If we introduce integral term (4.56) with regard to the controller output, as was often done in analogue controllers, i.e.

$$u_I(k-1) = u(k-1) = x(k) \qquad (4.68)$$

then, at the moment controller (4.57) is connected, there will be a step in output $u(k)$ corresponding to the size of control error $e(k)$ at that instant (in such a case it follows that it is advisable to connect the controller at a point where the control error is more or less zero).

We have tacitly assumed that set point w, fixed before the controller is connected, does not change at the instant of connection, that is that $w(k) = w(k-1)$.

The shortcoming of method (4.67) is that the control error is removed slowly after connection, whereas the disadvantage of (4.68) is that it results in step $u(k)$. Sometimes set point tracking is used in bumpless controller connection, where $w(j) = y(j)$ and therefore $e(j) = 0$, for $j = k-1, k$. The connection is bumpless for both (4.67) and (4.68), but has the disadvantage that, after connection, you must enter the real set point for the controlled variable. One possible solution is to switch between set points at the instant of connection, which can be implemented using a set point switch with two contacts. The controlled variable is connected to the first contact (for set point tracking) whereas the pre-fixed set point is connected to the second contact through a first-order lag or rate limiter. When the controller is switched from manual to automatic control, the switch instantaneously moves from the first to the second contact (i.e. $e(k-1) = 0$, but $e(k)$ is, already, generally nonzero). Controller (4.57) then behaves in the same way on connection as it does in operation and a change in the set point will occur.

Comparing position controller (4.56)–(4.57) with incremental controller (4.59)–(4.61) it follows that:

- The initial value for the integral term or bias u_I does not have to be calculated for an incremental controller. The change in controller output at the instant of connection is mainly determined by the difference between $e(k)$ and $e(k-1)$. As a result, the control error is compensated slowly just as in the case of a position controller which includes u_I according to (4.67). The solution is to switch set points at the instant of connection.

- Both controllers are equal in value if the following relation still holds:

$$u_I(k) = u(k-1) - r_0 e(k-1) + r_{-1} e(k) - r_1 [e(k-1) - e(k-2)] \quad (4.69)$$

Generally $u_I(k) \neq u(k-1)$. Equation (4.69) is valid only for the linear range of the controller, i.e. if the value of controller output $u(k)$ is not limited by upper or lower boundaries.

- The value of integral component u_I may only be limited at upper or lower boundaries for a position controller.

Note

If a measured value is to be used as the tracking signal it should first be filtered.

4.5.2 Limiting the Integral Term and Wind-up of Controller

Let the cascade control loop in Figure 4.12 be PID controlled with a slave (auxiliary) and master (main) controller (R_P, R_H). The plant S_P and S_H is contaminated by disturbances v^P and v^H causing bias of variables $y^S = u^{mH}$ and $y^P = u^{mH}$. Therefore the actuator y^S may be at a stop $(y^S = y^S_{min} o r y^S = y^S_{max})$ and y^P can still remain within the set (technological) band.

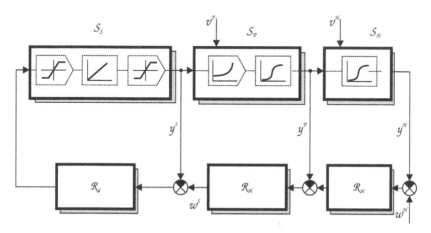

Figure 4.12. Cascade control loop

Assume that while the systems S_S and S_H have positive gain, system S_P has negative gain, as a result, the growth in y^S causes a drop in y^P and y^H.

Therefore, controller R_P has negative gain and controllers R_S and R_H positive gain.

If the actuator (drive, end element) comes to a stop and control error $e(k)$ is nonzero, then integral term $u_I(k)$ in the PID position controller starts to grow (theoretically) – see Equation (4.56); the increase of integral term $u_I(k)$ is stopped only after a change in polarity of $e(k)$. This results in computation of nonfeasible values for control output $u(k)$, causing the actuator to remain longer at a stop (the return to the control band depends on the relationship of the P, I and D terms). This delayed return of the actuator to the control band and consequent waste of energy supplied to the system can cause large overshoots in the process output, mainly in cascade control loops. This "over-excitation" of the controller is known as "wind-up" ("reset wind-up") or "integral saturation". It is caused by the failure to limit the integral term sufficiently after limiting the controller output in the actuator and/or in the final control element. The wind-up effect is demonstrated here in Figure 4.13, which shows the responses of a simple controlled system composed of a linear dynamic system and PI controller to set point changes. The manipulated variable is the controller output. The first step point change is a linear case without limitation of manipulated variable $u(k)$; in the second step point change the manipulated variable is limited but the integral term increases without limit. Compare the responses of $u(k), y(k)$.

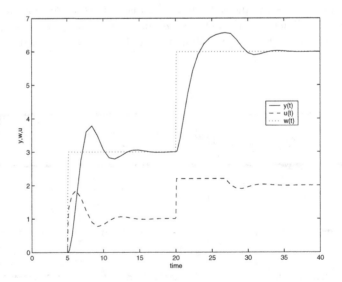

Figure 4.13. Wind-up effect

Some safeguards against the wind-up effect (anti wind-up algorithms) which must be based on the subsequent modification of the integral term are described below. These are effective for a controller in a simple control loop, but in a cascade control loop they are usually only useful for slave controller R_P, the limited output u^P of which corresponds to the limited value of actuator y^S. For the master controller, it is impossible to preset a constant output boundary which permanently corresponds to the stop in the control element (actuator). When dealing with the wind-up effect it is assumed that the master controller receives a signal (information) that the slave controller or actuator has reached a stop. Only Shinskey [63] and Fessl and Jarkovský [64, 65] (for PID as well as for LQ controllers) give solutions to cascade control in the references cited here. The other references do not treat this problem at all. Åström and Hägglund [9] recommend solving this problem by switching the master controller to manual control.

Glattfelder and Schauffelberger [66] and Rundquist [67, 68], among others, analyse wind-up from the point of view of nonlinear theory. Åström and Wittenmark [69] suggested one solution to the wind-up effect involving a state controller with observer. Rundquist [67, 68] examines several approaches and demonstrates that the different solutions can be regarded as variations on the state controller with observer.

It must be stated that the effectiveness of anti wind-up algorithms depends on whether the control device (actuator) has been limited as the result of disturbance or set point change, and also on the nature of the disturbance and the point at which it affects the system, as well as on the system dynamics (static or astatic, the ratio between time delay and the system's dominant time constant and the order of the system). It is therefore difficult to compare various anti wind-up algorithms and they all have different effects.

The controller output is limited to the control band

$$u \in \langle u_{\min}; u_{\max} \rangle \tag{4.70}$$

Which means that position controller Equation (4.56)–(4.57) and incremental controller Equation (4.59)–(4.61) must be supplemented by limitation (4.70). Due to varying modifications of the integral term, the output from incremental controller (4.59)–(4.61) and position controller (4.56)–(4.57) may differ after control element limiting. Incremental controller (4.59)–(4.60) with only output $\Delta u(k)$ implicitly includes a limited output because the calculated value is not put into effect if the control element or slave controller lies outside the control band. Since only the changes in the integral term and controller output $\Delta u(k)$ are calculated the wind-up effect does not occur.

Slave controller R_P is limited by the operating band of the final control element and thus the equation

$$u_{\min}^P \approx y_{\min}^S \qquad u_{\max}^P \approx y_{\max}^S \tag{4.71}$$

is valid.

A common solution to the wind-up effect was the use of tracking signal x for the integral term while limiting the controller output, where x denotes the controller's limited output. For the position controller described by (4.57), or more exactly by (4.56)–(4.58) and (4.70), a relation similar to (4.68) is employed, *i.e.* for slave controller R_P

$$u_I^P(k) = \begin{cases} u_{min}^P & \text{for } y^S = y_{min}^S \text{or } u(k) = u_{min}^P \\ u_{max}^P & \text{for } y^S = y_{max}^S \text{or } u(k) = u_{max}^P \end{cases} \tag{4.72}$$

Here the value of integral term u_I^P in fact jumps when it reaches stop y_{min}^S or y_{max}^S. Controller R_P and the actuator return back within the control band when control error e^P changes polarity.

A static limit on the integral term u_I on constant boundary

$$u_I \in \langle u_{I\,min}; u_{I\,max}\rangle \tag{4.73}$$

is analogous to Equation (4.72), where

$$u_{I\,min}^P \approx y_{min}^S \approx u_{min}^P \qquad u_{I\,max}^P \approx y_{max}^S \approx u_{max}^P \tag{4.74}$$

is generally used for controller R_P.

Controller R_P and the actuator return back within the control band when control error e^P changes polarity as in the previous case.

The effect of the static limit on the integral term u_I by (4.73) is shown on Figure 4.14 which shows responses of the same simple control loop as for Figure 4.13. The first step point change is linear without limitation of the manipulated variable $u(k)$; in the second step point change the manipulated variable is limited and the integral term is limited statically to a constant value as in (4.73). Compare the responses of $u(k)$, $y(k)$ with those of Figure 4.13.

Note
In some commercially available controllers the boundary of integral term u_I can be defined within a narrower band than that of the controller output. When the boundary between the integral term and the output has been reached, controller R_P and the actuator unit return to within band as soon as the control error e^P starts to fall.

The boundaries of u_I (4.73) cannot be introduced directly for incremental controller (4.59)–(4.61) and (4.70), where only the output limit (4.70) can be used. This is where the hitherto equivalent controllers (4.57) and (4.61) begin to differ. For an incremental controller, limit $u(k)$ in fact causes a step in the integral component of $usat(k) - u(k)$, where $usat(k)$ denotes the value limited in the band $\langle u_{min}; u_{max}\rangle$ and $u(k)$ stands for the calculated unlimited value. This leads to a subsequent recalculation of integral term u_I for position controller (4.57) to the value required to achieve limited values on the controller output (see Åström and Wittenmark [69], Clarke [70])

$$u_I(k) = u_I(k) + usat(k) - u(k) \tag{4.75}$$

where

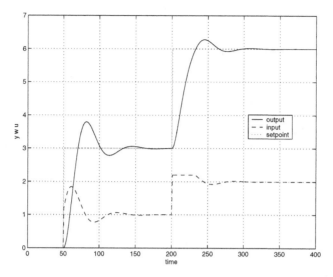

Figure 4.14. Static limit on the integral term u_I by (4.73)

$$usat(k) = \begin{array}{ll} u(k) & \text{for } u(k) \in \langle u_{min}; u_{max} \rangle \\ u_{min} & \text{for } u(k) < u_{min} \\ u_{max} & \text{for } u(k) > u_{max} \end{array} \qquad (4.76)$$

Controller R_P and the actuator return to the control band as soon as control error e^P drops. However this may cause the drive to oscillate greatly between stop and the control band and slow down the process of regulating disturbance and achieving zero control error (especially for the case of a larger drift in the measured values of the controlled variable).

Åström and Wittenmark [69] designed a solution to the wind-up effect for a state *controller with observer* as well as for a PID controller transformed into the state form. The solution is given below. First, the controller is transformed into the state form

$$\mathbf{x}(k+1) = \mathbf{F}\mathbf{x}(k) + \mathbf{G}y(k) \qquad (4.77)$$
$$u(k) = \mathbf{C}\mathbf{x}(k) + \mathbf{D}y(k) \qquad (4.78)$$

where $\mathbf{x}$ denotes the controller state, y the input (*i.e.* the control error) and u the controller output which is subject to a nonlinear type of limitation (*i.e.* by the drive) with output u^r. For the case where y and the limit of u have nonzero values, state $\mathbf{x}$, and therefore the calculated value of output u, will grow – unless the state value is made to correspond to the actual value of output u^r (estimate u^r or measured value u^m). Then, an explicit state observer is introduced so that, after multiplication by $\mathbf{K}$, Equation (4.78) can be added to Equation (4.77) and, instead of calculated value $u(k)$, the observed output value $u^r(k)$ is used.

This gives us

$$
\begin{aligned}
\mathbf{x}(k+1) &= \mathbf{F}\mathbf{x}(k) + \mathbf{G}y(k) + \mathbf{K}[u^r(k) - \mathbf{C}\mathbf{x}(k) - \mathbf{D}y(k)] \\
&= [\mathbf{F} - \mathbf{KC}]\mathbf{x}(k) + [\mathbf{G} - \mathbf{KD}]y(k) + \mathbf{K}u^r(k) \qquad (4.79) \\
&= \mathbf{F}_0\mathbf{x}(k) + \mathbf{G}_0 y(k) + \mathbf{K}u^r(k)
\end{aligned}
$$

If system (4.77) and (4.78) is observable, then matrix $\mathbf{K}$ can be chosen so that matrix $\mathbf{F}_0 = \mathbf{F} - \mathbf{KC}$ always has its eigenvalues inside the unit circle. The controller with observer then takes the form

$$
\begin{aligned}
\mathbf{x}(k+1) &= \mathbf{F}_0\mathbf{x}(k) + \mathbf{G}_0 y(k) + \mathbf{K}u^r(k) \\
u(k) &= \mathbf{C}\mathbf{x}(k) + \mathbf{D}y(k) \qquad (4.80) \\
u^r(k) &= sat[u(k)] = sat[\mathbf{C}\mathbf{x}(k) + \mathbf{D}y(k)]
\end{aligned}
$$

where sat denotes the limiter with boundaries $u_{\min}, u_{\max}$.

For a position controller, integral component u_I is the state x, $i.e.$ $x = u_I$ – see Equation (4.56) and (4.57). For $K = 1/T$

$$
\begin{aligned}
u_I(k) &= u_I(k-1) + r_{-1}e(k) \\
u(k) &= r_0 e(k) + u_I(k) + r_1[e(k) - e(k-1)] \\
u^r(k) &= sat[u(k)] \qquad (4.81)\\
u_I(k) &= u_I(k-1) + r_{-1}e(k) + \tfrac{1}{T_r}[u^r(k) - u(k)]
\end{aligned}
$$

where the final equation defines the re-calculation of the state-integral term. The relation (4.75) can be obtained for $K = 1$ (the so-called dead-beat observer).

Figure 4.15 shows a diagram (illustrating a continuous-time PID controller for the sake of clarity) where gain T_r (the observer constant) is introduced as feedback correction to the integrator state. The choice of value T_r can be used to influence the dynamics of the controller and its sensitivity to the noise resulting from measurements. The recommended value of T_r is proportional to T_I. The signal which is equal to $(u^r - u)/T_r$ can also be interpreted as the tracking signal from the integrator.

Hanus et $al.$ [71] suggest a recalculation of set point w^H to a value which leads to a limit on the control element. Another solution suggests a recalculation of set point w^H to a value which leads to a limit on the control element. This approach can be regarded as a variant on the controller with observer (see Rundquist [67, 68]). Like the previous (4.72)–(4.80), this solution can only be used directly for the subordinate controller R_P, since a boundary (4.70) or (4.73) for $u^H = w^P$, which permanently corresponds to drive boundary y^S, cannot usually be preset in the main controller R_H. The effects of disturbance

v^P can change the level of y^P so that w^P is within the boundaries, but y^S is limited.

In incremental controller (4.60), (4.61) it is possible to use the measured controller output $u(k-1)$ instead of $u^m(k)$, but, generally speaking, this is incorrect. Although wind-up disappears from both R_P and R_H controllers, it often leads to worse control within the control band. This is caused by neglected dynamic delay between $u^m(k)$ and $u(k-1)$ which may even lead to instability in the loop (see Fessl and Jarkovský [64, 65]). Measurement noise and disturbance acting on u^m interfere with the functioning of the controller.

It is, however, possible to use measured controller output u^m as the *tracking signal for main controller* R_H in a cascade control loop (if a stop of the actuator or slave controller has been reached), which means that

$$u_I^H(k) = y^P(k) = u^m(k) \qquad (4.82)$$

where the measured control output u^m is simultaneously the process out-

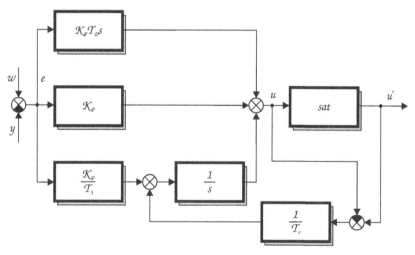

Figure 4.15. Block diagram of a PID controller where the wind-up effect is solved using Equation (4.81)

put y^P of the slave controller R_P, for which the set point is $u^H = w^P$. This solution, which removes or significantly suppresses wind-up in the main controller, is only described by Shinskey [65] and is generally little known among designers and control engineers. In incremental controller (4.60), (4.61) the corresponding value of $u(k-1)$ is calculated according to (4.69).

The *dynamic limitation of the integral term* can be used for both R_P and R_H. At the moment of limiting the control element, the integral term is "frozen", *i.e.* it is limited to a previously unknown dynamically variable boundary

$$*u_I = u_I(k) = u_I(k-1) \tag{4.83}$$

or, for incremental controller (4.60), during the "freeze" period

$$\Delta u_I(k) = 0 \tag{4.84}$$

The drive returns to the control band when terms P and D start to drop (this is slower than recalculation (4.75) when it is used for slave controller R_P, but usually better because the actuator does not oscillate so much around the stop).

Both (4.82) and (4.83) are the only methods given so far to remove or significantly diminish the wind-up effect in the main controller R_H in a cascade control loop. Figure 4.16 presents responses of the cascade control loop, shown in Figure 4.12, with disturbances and constant set point; the integral term is limited: (a) statically to a constant value as in (4.73), which is not too successful in cascade control; (b) dynamically according to (4.83). Compare the responses of main controller output $w^P = u^H$ and the impact on controlled variable y^H after limitation of the actuator.

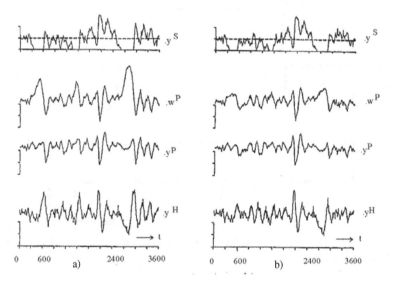

Figure 4.16. Limiting the integral term: (a) statically as in (4.73), (b) dynamically as in (4.83) in cascade control

The "frozen" value $*u_I$ (4.83) can also be taken as the tracking signal for the integration channel. The disadvantage of (4.82) and (4.83) is that the size of $u_I^H(k)$, $*u_I$ and the behaviour of y^P depend on the disturbance behaviour causing the actuator to reach the limit. Therefore, in some situations Equation (4.83), in others (4.82) results in smaller overshoot of the controlled variable, depending on the previously unknown disturbance behaviour. This shortcoming can be removed if the tracking signal is determined by a logical selector, which selects between the two alternative tracking signals to give the lowest process output overshoot.

Correcting the calculated manipulated variable.

The calculated controller output (manipulated variable) $u(k)$ can be dynamically limited using a so-called manipulated variable corrector (Fessl and Jarkovský [64, 65]), which respects the true values of measured controller output u^m. The corrector, originally developed for an adaptive controller with recursive identification, can also be used for PID and similar controllers. In incremental controller (4.60), (4.61) the last calculated value of $u(k-1)$ is replaced by corrected value $u^K(k)$. Denoting v the difference between the computed and measured value of u

$$v = u(k-1) - u^m(k)$$

then, it yields

$$u(k-1) = u^K(k) = sat_{e\,\max}(v) + u^m(k) \tag{4.85}$$

where $e\,\max > 0$ is the maximum permitted control error (over several sampling periods) in the relevant control loop, and sat denotes the limiter in boundaries $-e\,\max$ and $+e\,\max$. The value of $e\,\max$ must also be chosen with regard to possible changes in controlled variable $y(k)$ and manipulated variable $u(k)$ when disturbance is great or there are large changes in the set point. Equation $u^K(k) = u(k-1)$, i.e. standard incremental controller (4.61), remains valid when $|v| < e\,\max$. For the unsuitable limit $e\,\max = 0$, the controller changes to one using the constantly measured manipulated variable u^m. In contrast to the dynamic limitation on integral term (4.83), the effect of the final control element's static nonlinearity is now compensated as well. The value of u_I can subsequently be recalculated for position controller (4.56)–(4.57) as

$$u_I(k-1) = u_I(k-1) + u^k(k) - u(k-1) \tag{4.86}$$

The application of Equation (4.85) is often reasonable combined with another condition (especially the actuator position y^S near the real stop, i.e. apply corrector for $y^S < 3\%$ or $y^S > 97\%$, for example). Unlike constant boundary usat (4.76) used for recalculating value $u_I(k)$ employing (4.75), the "boundary" given by u^K is variable and obtained from the true measured value of u^m. Corrector (4.85) is particularly useful for main controller R_H and is an alternative solution to the use of tracking signal (4.82) and dynamic limitation on the integral term (4.83).

Rundquist [67, 68] compares several anti wind-up algorithms for a simple control loop, and Šulc [72] describes real hydraulic and pneumatic controllers and suitable solutions.

Apart from using (algorithmic) controller output limitation (4.70), output changes are often limited by

$$\Delta u(k) \in \langle \Delta u_{\min}; \Delta u_{\max} \rangle \tag{4.87}$$

and also the values of controller output $u(k)$ in relation to another variable $z(k)$

$$u(k) = sat_z[u(k) - z(k)] + z(k) \tag{4.88}$$

The integral term is changed indirectly by limiting the controller output, whereas in incremental controller (4.61) we in fact alter the increment of the integral term $\Delta u_I(k)$, and consequently integral term $u_I(k)$. It follows that boundary $u_{\min}^H, u_{\max}^H$ can be changed directly rather than altering the value of integral component $u_I(k)$ in the main controller of a cascade control loop. When the slave controller reaches the output boundary, the boundary of the main controller can be defined to equal the current value of its output u^H, and consequently its integral term will be corrected. For example, the following algorithm can be used (assuming the same range of values for u^H and u^P, *e.g.* 0, 100% and the same polarity for the controllers)

$$u_{\max}^H = u^H + u_{\max}^P - u^P \qquad u_{\min}^H = u^H + u_{\min}^P - u^P \tag{4.89}$$

The wind-up effect can also result from the use of selection circuit (for example a circuit to select a minimum or maximum) to switch over signals ("override control"). Consider the structure of a control loop in which the control signal for the slave controller or control element is selected from the outputs of two or more main controllers. Then, if the "unselected" PID controllers remain active their integral terms will change even if the output is not put into effect. One answer to this problem is to set the output from unselected controllers to the value of the output from the selected controller. Alternatively, we can constantly change the boundaries of these controllers.

Note
Except for corrector (4.85), all algorithms for limitation of the integral component are based on the assumption that the stop of the final control element (actuator) and the boundary of the relevant controller correspond. If, due to bad (mechanical) tuning, the actuator reaches the stop before the controller does, the controller will continue the integration process, resulting in wind-up, *i.e.* greater integration of the controller and consequently greater overshoot in the process output. Unfortunately even the use of digital output to signal the actuator boundary is not entirely reliable.

"The control paradox": sometimes a seemingly inexplicable phenomenon occurs. Even though there may be a large control error, the control element starts to move in the wrong direction at the stop, closes up rather than opening, for example, or perhaps oscillates near the stop. This is caused by an unsuitable anti wind-up algorithm, drift in the process output, and possibly also by an unsuitably large integral time constant.

4.5.3 Limited Precision in Calculation

The precision of the A/D and D/A converters used (quantization level), computer word length together with related errors in rounding off and the use of fixed or floating point arithmetic, all affect the performance of a digital controller. For details see Isermann [54] and Åström and Wittenmark [69]. All the conditions above also have an effect on the choice of sampling period and controller parameters K_P, T_I and T_D (see Section 4.4).

An appropriate normalization of variables (for example subtracting the reference values from the controlled and manipulated variable) increases the number of valid places for the calculated values and therefore improves the precision of the calculation. Limited precision causes bias – a permanent control error or limiting cycles. This will be illustrated by an example. The change in the integral component is

$$\Delta u_I(k) = \frac{K_P T_0}{T_I} e(k) \qquad (4.90)$$

If we use a 12 bit converter with quantization $1/4096$, and $K_P T_0 = 1$ and $T_I = 3600$, error not exceeding 88% of the converter band does not cause the change in Δu_I to be greater than the level of quantization. No change occurs when the common 8 bit converter with a level of $1/256$ is used. The answer is to double the length of the word to calculate the integral component or round up all calculations, or add to the random component (noise on the level of quantization) together with suitable normalization of the variables – Bristol [73]. Similarly, the derivative term is either invalid, or else is valid with steps

$$\Delta u_D(k) = [e(k) - e(k-1)] \frac{K_P T_D}{T_0} \qquad (4.91)$$

It is clear from the above that there is a difference in behaviour between incremental controller (4.59), (4.60), which only calculates the variously rounded off values of Δu_I, and position controller (4.56), (4.57). Another difference (apart from the effect of limited controller output – see Section 4.5.2) is in the behaviour towards changes in the parameters and in set point w.

4.5.4 Filtering the Measured Variables

Sampling using a period T_0 and frequency $\omega_0 = 2\pi/T_0$ creates a low frequency signal with differential frequency $\omega_2 = |\omega_0 - \omega_1|$ and unchanged amplitude from signal $x(t)$ with higher frequency ω_1, when $\omega_1 \geq \omega_0/2$ and period $T_1 < 2T_0$. This is illustrated in Figure 4.17.

This *"aliasing effect"* would result in an unsuitable calculation of the controller output. These higher frequencies must, therefore, be filtered through a low frequency filter before being used in a controller. A filter also limits the effects of the acting measurement noise, *etc*. The most commonly used

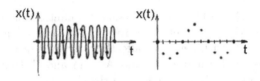

Figure 4.17. Signal $x(t)$ with higher frequency ω_1 sampled with frequency $\omega_0 \ll \omega_1$

are analogue filters (first or higher order lag) or specialized filters (the Butterworth, Bessel or Chebyshev filter). Isermann [54] and others deal with the problems of filters in more detail. Analogue filters are more difficult to apply to larger sampling periods and are therefore replaced by digital filters. An analogue signal can be sampled with digital filter period $T_f = T_0/m < T_0$, where the filter output is used with period T_0. Effective filters for multiple sampling based on the on-line identification approach have been designed, which filter the signal better than, for example, the commonly used method of simply taking an average of the latest values. However, filtration not only suppresses noise but the useful signal as well. For this reason it was suggested use be made of a digital correction term with inverse transfer function F^{-1} to the transfer function of an analogue first-order filter F

$$F(z) = \frac{(1-a)z}{z-a} \qquad a = \exp(-T_0/T_f) \tag{4.92}$$

The input into the analogue filter is a signal affected by noise $y_n(k)$, the useful component $y(k)$ of which is calculated with period T_0. Using (4.92) it follows that

$$y(k) = \frac{y_f(k) - ay_f(k-1)}{1-a} \tag{4.93}$$

where $y_f(k)$ denotes the output from the analogue filter.

Special filters are used for signals affected by abnormal noise distribution. Clarke [70] describes a filter which reliably removes noise having short-term rectangular pulses ("peaks"). This algorithm, where parameter gapmin is the minimal value of variable $gap(k)$, $x(k)$ and $y(k)$ denotes filter input and output, resp., follows:

```
      d(k)= x(k) - y(k-1)
if    abs(d(k)) < gap(k-1)
then y(k) = x(k)
      gap(k)= max (gap(k-1)*0.5, gapmin)
else  y(k) = y(k-1) + gap(k)*sign(d(k))
      gap(k)= gap(k-1)*2
```

end

4.5.5 Industrial PID Controllers

The development of electronics and particularly computer technology (micro-processors) has meant that practically all commercial electronic controllers are digital and no longer contain analogue elements as they did before. PID controllers exist as one or multi-control loop controllers or as program (function) blocks in PLCs or control systems.

It should be noted that the output from PID-type controllers either takes the form of analogue output (AO) or digital output (DO) – one digital output is usually used (for switching on and off) or two digital outputs (for switching on and off or to control the actuator in the "increasing" and "decreasing" direction, *etc.*). Apart from these main outputs there may be other auxiliary digital outputs for various indications *etc.* Furthermore, controllers are commonly connected to the communication interface RS422/485 or other company channels. An analogue output with period T_0 generates an amplitude-modulated impulse signal, whereas a digital output produces a pulse width-modulated signal, *i.e.* a signal with a width proportional to a calculated value $u(k)$. This makes a digital PID controller with digital outputs rather similar to the classic two or three-position controller. For a controller with digital outputs, it is wise to use an incremental controller with output $\Delta u(k)$ – Equation (4.13b) or (4.59)–(4.60); the size (and polarity) of the change $\Delta u(k)$ is introduced directly into the relevant digital output. In the case of a controller with an analogue output, it is possible to use either algorithm (4.16) or (4.59)–(4.61), alternatively the component form (4.27)–(4.28) or (4.56)–(4.57).

In addition to its own parameters (K_P, T_I, and T_D, plus the T_f parameter for the D-term filter), a PID controller block also has upper and lower output boundaries (minimum and maximum values) $u_{\min}$, $u_{\max}$ and sampling period T_0 (which can be defined as the period of the whole control program). It is sometimes also possible to include the greatest permitted change $+\Delta u_{\max}$, $-\Delta u_{\max}$.

The polarity of the controller is set according to the polarity of the gain of the controlled system – terms indirect or reverse action controller and direct controller are used. These names are taken from the direction in which the manipulated variable operates in relation to the direction of changes in the controlled variable. If the controlled system has negative gain then an increase in the controlled variable results in a direct increase of the P term causing a direct increase in the manipulated variable.

The controller block often includes feed-forward signals and possibilities to use proportional or cascade control. Functions for switching between manual and automatic control and manual setting of the value for controller output are standard. Many industrial controllers can be programmed: the set point is generated by a program (for programmable control). In the case of so-called multi-function controllers, the user can even program the structure of the control loop from a library of function blocks. Most companies manufacturing controllers now supply them with an auto-tuning function which is (a one-shot) function to tune the controller parameters automatically. Some companies developing algorithms for auto-tuning offer fuzzy logic algorithms. Auto-tuning algorithms are quite often based on evaluation of the step-response in both open and closed loops. It is tacitly assumed that only the manipulated variable acts on the controlled system, which is a single-variable one. This means that the algorithms fail to consider the effects of various measured and unmeasured disturbances and nonlinearities. Therefore these algorithms (and the controllers which use them to tune their parameters) tend to work better for set point transitions and worse for a disturbance-contaminated loop.

4.6 Survey of Self-tuning PID Controllers

The structure of these controllers must be designed so that the numerator of the discrete transfer function is always in the form of a second-order polynomial. The form that is most frequently used is given below:

$$G_R(z) = \frac{U(z)}{E(z)} = \frac{q_0 + q_1 z^{-1} + q_2 z^{-2}}{1 - z^{-1}} \tag{4.94}$$

The structure of discrete transfer function (4.94) leads to the incremental equation of digital PID controller (4.14), where parameters q_0, q_1 and q_2 depend on the type of discretization of the integral component.

A suitable version of the discrete transfer function of a digital PID controller to apply the pole assignment method is the structure

$$G_R(z) = \frac{q_0 + q_1 z^{-1} + q_2 z^{-2}}{(1 - z^{-1})(1 + \gamma z^{-1})} \tag{4.95}$$

By choosing one of the above or similar structures we ensure that the parameters of the digital PID controller correspond to the P, I, and D components of its continuous-time version. The algorithm of this controller is composed of the parameter estimates of process model $\hat{\Theta}(k)$. These are then used to calculate the optimal parameters of the digital PID controller and consequently controller output $u(k)$ at each sampling period.

In this section we will give a clear picture of some work taken from texts dealing with the design of self-tuning PID controllers, *i.e.* adaptive PID controllers with recursive identification.

Corripio and Tomkins [74] designed an algorithm based on estimation of the parameters of a second-order model using recursive identification applied by the instrumental variable method. The parameter estimates $\hat{a}_1, \hat{b}_1$ and $\hat{a}_2$ (assuming $\hat{b}_2 = 0$) are used to calculate the controller parameters and included in those relations obtained by Dahlin [75, 76].

Wittenmark [77] analyzed digital PID controllers designed using the pole assignment method together with the recursive least squares method to estimate the process model parameters. Six different digital PID controller structures are discussed in relation to a more generalized control algorithm and there is a particular comparison between the suggested algorithm and the common version based on the pole assignment method. The advantages and disadvantages of each individual controller are demonstrated in simulated examples.

A survey and discussion of several self-tuning PID controller types is given by Ortega and Kelly [78]. They treat two types of explicit controllers and one implicit, all based on recursive identification and the pole assignment method. They also include a controller which minimizes the explicit criterion.

Kim and Choi [79, 80] modified the algorithm of the implicit controller given in [78], limiting large oscillations of the controlled output due to set point changes.

Kofahl and Isermann [81] developed a tuning procedure based on the algebraic calculation of critical gain and critical periods of oscillation, followed by application of the Ziegler–Nichols method.

Radke and Isermann [82] showed a self-tuning PID controller built on the estimation of process model parameters using the recursive least squares method and numeric optimization of parameters based on quadratic criteria. If $\mathbf{q}^T = [q_0, q_1, q_2]$ is the vector of the parameters of the controller (4.94), then the most generalized method of determining their optimal values is to minimize quadratic criteria

$$J = \sum_{k=0}^{N} [e^2(k) + rK_S^2 \Delta u^2(k)] \tag{4.96}$$

where $e(k)$ is the tracking error, $\Delta u(k) = u(k) - u(k-1)$ is the controller output error, K_S is the process gain, and r is the weighting factor of the controller output. The optimal controller parameters are obtained by solving equation

$$\frac{\partial J}{\partial q_i} = 0; \quad \text{for } i = 1, 2, 3 \tag{4.97}$$

The parameter optimization above can only be performed analytically for lower order controllers or systems. In general, optimization must be carried out numerically. Two ways are given to solve criterion (4.97) in reference [82]; either in the time field by solving the differential equation of a closed control loop with suitable external excitement, or in the z domain using the Parseval integral. Optimization of parameters according to relation (4.97) is carried out numerically.

Tzafestas and Kapsiotis [83] designed a combined self-tuning controller based on the pole assignment method followed by the optimization of the vector of the controller parameters using criterion (4.96).

Gawthrop [84] used a hybrid self-tuning controller to tune an external PID controller. The implementation of this kind of control system, however, presumes very high integration of the hybrid controller using special loops, and produces a controller which is part digital and part pseudo-analogue.

Neuman [85] designed a self-tuning predictive PI controller (PIR) derived from the algorithm of the classic incremental PI controller. The predicted output value can be used either in the integral or proportional component, or even in both. This controller permits the use of both one-step and higher order predictors. The recursive parameter estimate of the linear predictor is performed using the recursive least squares method. This is, then, a controller with fixed parameters, adaptive in that it uses the predicted output value to determine the control output.

Alexík [86] describes analytical PID algorithm which has adaptive self-tuning performance. Its synthesis is performed in the continuous-time domain. This means that identification is followed by recalculation into the continuous-time domain and the final gain of the algorithm is compensated due to the sampling period. Bányász et al. [87, 88] suggested several modifications of the self-tuning PID controller. Again, the least squares method is used to estimate parameters of the process model and very simple relations have been deduced to design the controller. The algorithms are also suitable for the control of processes which are already known or have unknown time delay.

Some modifications of PID type controllers have been developed by Böhm et al. [89]. Each modification differs from the others in the choice of model structure; controller design is solved on the basis of minimizing the quadratic criterion using the polynomial method.

Ziegler–Nichols' method is very often used as the first method for tuning a control loop in practice. It has some advantages and some disadvantages. Its simplicity and universality are advantages and the low accuracy is the basic disadvantage. Controlled circuits tuned by this method gave relatively high values of overshoot (10–60%). The dynamic inversion method (see Vítečková et al. [90, 91]) retains the simplicity of the Ziegler–Nichols method but it is more accurate and universal. However, the method described thereinafter is suitable only for some types of controlled systems which are controlled by some types of PID controllers. The transfer functions of the controlled systems are given in the s complex plane there, however, they can be easily transformed into the z complex plane and in the STC algorithm are used as discrete ones. However, there are several methods of reducing the order of the controlled system or it can be estimated directly into a suitable one using recursive identification, i.e. RLSM (recursive least squares method).

A self-tuning PID controller algorithm based on this approach has been derived by Bobál et al. [92]. The process was identified using the regression (ARX) continuous-time model, the RLSM with directional forgetting was applied. The recursive parameter estimates of the continuous-time model (differential equation) have been used for synthesis of the PID controller.

Algorithms for self-tuning PID controllers based on the use of the Ziegler–Nichols criterion, modified for digital loops, have been designed by Bobál [93]. Synthesis of the controller is based on determination of critical proportional gain and critical period of oscillation when the closed loop is at the stability boundary. The critical parameters are calculated from the recursive parameter estimates of the process model without the necessity of using some test signal to make the control loop oscillate at the stability boundary. The recursive least squares method with directional forgetting is used in the identification procedure [49, 50]. A control algorithm has been derived for second-order models [94] together with a controller for third-order systems [33, 95], as well as for general n-th order systems [59, 96]. The approach given can easily be modified to enable the derivation of algorithms to control systems with time delay [97, 98].

4.7 Selected Algorithms for Self-tuning PID Controllers

In this chapter certain explicit self-tuning controllers will be presented which have been algorithmically modified in the form of mathematical relations or as flow diagrams so as to make them easy to program and apply. All the algorithms given below are included in the MATLAB® Toolbox. Some are original algorithms based on a modified Ziegler–Nichols criterion, others have been culled from publications and adapted to make them more accessible to the user.

4.7.1 Dahlin PID Controller

The algorithm of this controller, which is given in [74], calculates the controller output using an incremental form of relation discretized by the forward rectangular method (4.16)

$$
\begin{aligned}
u(k) = K_P \Big\{ &e(k) - e(k-1) + \frac{T_0}{T_I} e(k) \\
&+ \frac{T_D}{T_0} [e(k) - 2e(k-1) + e(k-2)] \Big\} + u(k-1)
\end{aligned}
\tag{4.98}
$$

This controller uses parameter estimation vector (3.14) in the form

$$
\hat{\Theta}^T(k) = [\hat{a}_1, \hat{a}_2, \hat{b}_1]
\tag{4.99}
$$

and since $\hat{b}_2 = 0$, regression vector (3.15) is

$$
\Theta^T(k-1) = [-y(k-1), -y(k-2), u(k-1)]
\tag{4.100}
$$

The following relations were obtained for individual controller parameters:

$$K_P = -\frac{(a_1 + 2a_2)Q}{b_1} \qquad T_I = -\frac{T_0}{\frac{1}{a_1+2a_2} + 1 + \frac{T_D}{T_0}} \qquad T_D = \frac{T_0 a_2 Q}{K_P b_1} \qquad (4.101)$$

Variable Q in Equation (4.101) is defined by the relation

$$Q = 1 - e^{-\frac{T_0}{B}} \qquad (4.102)$$

where B is known as the adjustment factor which characterizes the dominant time constant of the transfer function according to changes made to the process output of a closed control loop. The smaller the value of B, the faster the response of the closed control loop.

To avoid oscillation in the process output variable it is advisable to choose the initial control parameter estimates using the following relations. These are basically the inverse expression of Equation (4.101).

$$C = 1 + \frac{T_0}{T_I} + \frac{T_D}{T_0}$$

$$a_1 = -\frac{1 + 2\frac{T_D}{T_0}}{C} \qquad a_2 = \frac{T_D}{T_0 C} \qquad b_1 = \frac{Q}{K_P C} \qquad (4.103)$$

Before devising our own algorithm we must take its numerical stability into account. It follows from Equation (4.101) that if $b_1 = 0$ division by zero may occur. Therefore when we start the algorithm we must avoid choosing zero for parameter b_1 and manipulate the algorithm bearing this danger in mind.

4.7.2 Bányász and Keviczky PID Controller

This controller has been derived from and analysed in references [87, 99, 100, 88]. The PID controller discrete transfer function is considered in its standard form

$$G_R(z) = \frac{Q(z^{-1})}{P(z^{-1})} = \frac{q_0 + q_1 z^{-1} + q_2 z^{-2}}{1 - z^{-1}} \qquad (4.104)$$

Digital filter G_F is serially connected to the controller. The controlled process is described by the discrete transfer function

$$G_P(z) = \frac{B(z^{-1})}{A(z^{-1})} = \frac{b_0 + b_1 z^{-1}}{1 + a_1 z^{-1} + a_2 z^{-2}} z^{-d}$$
$$= \frac{b_0(1 + \gamma z^{-1})}{1 + a_1 z^{-1} + a_2 z^{-2}} z^{-d} \qquad (4.105)$$

where $b_0 \neq 0$ and $d > 0$ is the number of time delay steps. Figure 4.18 shows a block diagram of the closed loop. The controller polynomial $Q(z^{-1})$ is chosen so as to validate

$$Q(z^{-1}) = q_0(1 + q_1' z^{-1} + q_2' z^{-2}) = q_0(1 + a_1 z^{-1} + a_2 z^{-2}) = q_0 A(z^{-1}) \quad (4.106)$$

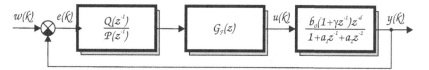

Figure 4.18. Block diagram of the control loop for a Bányász and Keviczky controller

which means that

$$q_1' = \frac{q_1}{q_0} \qquad q_2' = \frac{q_2}{q_0} \tag{4.107}$$

This idea allows us to simplify the control loop of Figure 4.18 to that of Figure 4.19 where the integrator and pure time delay are directly connected in series when relation (4.108) for integrator gain is valid.

$$k_I = q_0 b_0 \tag{4.108}$$

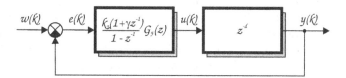

Figure 4.19. Simplified block diagram of a control loop

The correction filter can be used to compensate for unwanted interference caused by the expression $1 + \gamma z^{-1}$ as well as for other purposes during controller synthesis. This controller assumes knowledge of the number of time delay steps. Parameter estimates vector (3.15) has the form

$$\hat{\Theta}^T(k) = [\hat{a}_1, \hat{a}_2, \hat{b}_0, \hat{b}_1] \tag{4.109}$$

and regression vector (3.16) is

$$\phi^T(k-1) = [-y(k-1), -y(k-2), u(k-d), u(k-d-1)] \tag{4.110}$$

The model parameter estimates are then used to calculate the controller parameters according to the simple relation (3.15)

$$\gamma = \frac{b_1}{b_0} \qquad q_0 = \frac{k_I}{b_0} \qquad q_1 = q_0 a_1 = \frac{k_I}{b_0} a_1 \qquad q_2 = q_0 a_2 = \frac{k_I}{b_0} a_2 \tag{4.111}$$

where

$$k_I = \frac{1}{2d-1} \quad \text{for } \gamma = 0 \tag{4.112a}$$

$$k_I = \frac{1}{2d(1+\gamma)(1-\gamma)} \quad \text{for } \gamma > 0 \tag{4.112b}$$

If $\gamma < 1$, Equation (4.112a) is used together with a serially connected digital filter

$$G_F(z) = \frac{1}{1+\gamma z^{-1}} \tag{4.113}$$

Relations (4.111) and (4.112) are then inserted into the equation to calculate the controller output

$$u(k) = q_0 e(k) + q_1 e(k-1) + q_2 e(k-2) + u(k-1) \tag{4.114}$$

If $b_0 = 0$ division by zero may occur.

4.7.3 Digital PID Controllers Based on the Pole Assignment Method

A controller based on the assignment of poles in a closed feedback control loop is designed to stabilize the closed loop while the characteristic polynomial should have previously determined poles. Apart from the stability requirement, good poles configuration can make it relatively easy to obtain desired closed loop response (*e.g.*, the maximum overshoot, damping, *etc.*). The general approach to this method on the basis of algebraic theory is examined in detail in Chapter 5.

A PID controller design which ensures the required control loop dynamic behaviour by choosing the characteristic polynomial is given in references in connection with various loop block structures [77, 78].

Structure of the PID–A Control Loop

This controller design comes from the general closed loop block diagram shown in Figure 4.20, where

$$G_P(z) = \frac{Y(z)}{U(z)} = \frac{B(z^{-1})}{A(z^{-1})} \tag{4.115}$$

is the discrete transfer function of the controlled plant with polynomials

$$A(z^{-1}) = 1 + a_1 z^{-1} + a_2 z^{-2} \qquad B(z^{-1}) = b_1 z^{-1} + b_2 z^{-2} \tag{4.116}$$

$$G_R(z) = \frac{U(z)}{E(z)} = \frac{Q(z^{-1})}{P(z^{-1})} \tag{4.117}$$

is the transfer function of a controller with polynomials

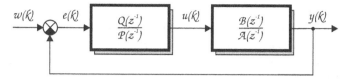

Figure 4.20. Block diagram of control loop for a PID–A controller

$$P(z^{-1}) = (1 - z^{-1})(1 + \gamma z^{-1}) \qquad Q(z^{-1}) = q_0 + q_1 z^{-1} + q_2 z^{-2} \quad (4.118)$$

From Equation (4.117) it is possible to determine the controller equation in the form

$$U(z) = \frac{Q(z^{-1})}{P(z^{-1})} E(z) \qquad (4.119)$$

and by inserting polynomials (4.118) into Equation (4.119), the relation to calculate the controller output becomes

$$u(k) = q_0 e(k) + q_1 e(k-1) + q_2 e(k-2) + (1 - \gamma)u(k-1) + \gamma u(k-2) \quad (4.120)$$

Further, the following relation can be obtained for the control transfer function of the closed loop shown in Figure 4.20

$$G_W(z) = \frac{Y(z)}{W(z)} = \frac{B(z^{-1})Q(z^{-1})}{A(z^{-1})P(z^{-1}) + B(z^{-1})Q(z^{-1})} \qquad (4.121)$$

where the characteristic polynomial is to be found in the denominator of (4.121). By choosing the characteristic polynomial

$$D(z^{-1}) = 1 + \sum_{i=1}^{n_d} d_i z^{-i}, \quad n_d \le 4 \qquad (4.122)$$

in the polynomial equation

$$A(z^{-1})P(z^{-1}) + B(z^{-1})Q(z^{-1}) = D(z^{-1}) \qquad (4.123)$$

we fix the desired pole placement for the transfer function (4.121). This is achieved by selecting the correct parameters for controller polynomials (4.119) which are the solution to polynomial Equation (4.123). The characteristic polynomial (4.123) can be defined by various methods. Most frequently used are those which meet the following requirements:

- the response of a continuous-time second-order plant;
- the response of a discrete second-order plant;
- dead-beat control;
- quadratic optimal control;
- desired response according to the user.

PID–A1 Controller

This type of a controller was derived using method 1 where the required control response of a closed loop can be achieved by selecting natural frequency ω_n and damping factor ξ in the characteristic equation for a continuous-time second-order plant

$$s^2 + 2\xi\omega_n s + \omega_n^2 = 0 \tag{4.124}$$

If the polynomial $D(z^{-1})$ is chosen in the form

$$D(z^{-1}) = 1 + d_1 z^{-1} + d_2 z^{-2} \tag{4.125}$$

then the following relations to calculate the coefficients for a sampling period T_0 can be derived:

$$d_1 = -2\exp(-\xi\omega_n T_0)\cos(\omega_n T_0\sqrt{1-\xi^2}); \quad \text{for } \xi \leq 1$$
$$d_1 = -2\exp(-\xi\omega_n T_0)\cosh(\omega_n T_0\sqrt{1-\xi^2}); \quad \text{for } \xi > 1 \tag{4.126}$$
$$d_2 = \exp(-2\xi\omega_n T_0)$$

For polynomial (4.124) to have stable poles $\xi > 0, \omega_n > 0$ must be valid. The damping factor ξ is chosen according to the requirement of an oscillating or non-oscillating control response. Wittenmark and Åström [101] recommend choosing a value for natural frequency which makes the inequality $0.45 \leq \omega_n T_0 \leq 0.90$ valid.

If polynomial (4.125) is inserted into the right side of Equation (4.123), a series of four algebraical linear equations for four unknown parameters is obtained which can be written in the matrix form

$$\begin{bmatrix} b_1 & 0 & 0 & 1 \\ b_2 & b_1 & 0 & a_1 - 1 \\ 0 & b_2 & b_1 & a_2 - a_1 \\ 0 & 0 & b_2 & -a_2 \end{bmatrix} \begin{bmatrix} q_0 \\ q_1 \\ q_2 \\ \gamma \end{bmatrix} = \begin{bmatrix} x_1 \\ x_2 \\ x_3 \\ x_4 \end{bmatrix} \tag{4.127}$$

The first matrix on the left side of (4.127) depends only on the parameters of the controlled system, the next vector contains the unknown parameters of the controller, being the solution to the system of equations (4.127), and the vector on the right depends on the number of poles n_d and their position in the z complex plane. In this case the elements in the vector are given by the relations

$$x_1 = d_1 + 1 - a_1 \qquad x_2 = d_2 + a_1 - a_2 \qquad x_3 = a_2 \qquad x_4 = 0 \tag{4.128}$$

The solution of the equation system, *i.e.* the relations to calculate the controller parameters, are is given in Table 4.3.

PID–A2 Controller

When controlling technological processes, in many cases there is a requirement for a control response without or only with limited overshoot. Then it is a good idea to choose the characteristic polynomial in the form [93, 102]

$$D(z) = (z - \alpha)^2 [z - (\alpha + j\omega)][z - (\alpha - j\omega)] \qquad (4.129)$$

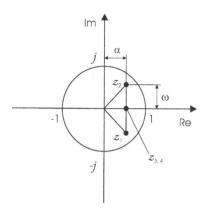

Figure 4.21. Pole assignment in polynomial $D(z)$

Table 4.3. Relations to calculate PID–A controller parameters

Controller parameters	PID–A1	PID–A2
q_0	$\dfrac{1}{b_1}(d_1 + 1 - a_1 - \gamma)$	$\dfrac{r_2 - r_3}{r_1}$
q_1	$\dfrac{a_2}{b_2} - q_2(\dfrac{b_1}{b_2} - \dfrac{a_1}{a_2} + 1)$	$-\dfrac{r_4 + r_5}{r_1}$
q_2	$-\dfrac{s_1}{r_1}$	$\dfrac{x_4 + \gamma a_2}{b_2}$
γ	$q_2 \dfrac{b_2}{a_2}$	$\dfrac{r_6}{r_1}$

Characteristic polynomial (4.129) has a pair of complex conjugated poles $z_{1,2} = \alpha \pm j\omega$ placed inside the unit circle at interval $0 \le \alpha < 1$ and double real pole $z_{3,4} = \alpha$, where $\alpha^2 + \omega^2 < 1$ (see Figure 4.21). The parameter α can be used to change the speed of the control response and the size of the changes in the controller output at the same time. It is also possible to change parameter ω to select different overshoots. The left side of the equation system in this controller is similar to that of system (4.127), the vector elements on the right side are given by relations

$$x_1 = c + 1 - a_1 \qquad x_2 = d + a_1 - a_2 \qquad x_3 = f + a_2 \qquad x_4 = g \qquad (4.130)$$

where

$$c = -4\alpha \qquad d = 6\alpha^2 + \omega^2$$
$$f = -2\alpha(2\alpha^2 + \omega^2) \qquad g = \alpha^2(\alpha^2 + \omega^2) \qquad (4.131)$$

The relations for calculating the controller parameters are again given in Table 4.3, while the relations to calculate the auxiliary variables needed to determine the controller parameters take the form

$$s_1 = a_2[(b_1 + b_2)(a_1b_2 - a_2b_1) + b_2(b_1d_2 - b_2d_1 - b_2)]$$
$$r_1 = (b_1 + b_2)(a_1b_1b_2 + a_2b_1^2 + b_2^2)$$
$$r_2 = x_1(b_1 + b_2)(a_1b_2 - a_2b_1)$$
$$r_3 = b_1^2x_4 - b_2[b_1x_3 - b_2(x_1 + x_2)] \qquad (4.132)$$
$$r_4 = a_1[b_1^2x_4 + b_2^2x_1 - b_1b_2(x_2 + x_3)]$$
$$r_5 = (b_1 + b_2)[a_2(b_1x_2 - b_2x_1) - b_1x_4 + b_2x_3]$$
$$r_6 = b_1(b_1^2x_4 - b_1b_2x_3 + b_2^2x_2) - b_2^3x_1$$

Example 4.1. Consider three controlled processes described by the following continuous-time transfer functions:
(a) Stable $G_A(s) = \frac{1}{(5s+1)(10s+1)}$
(b) With nonminimum phase $G_B(s) = \frac{1-4s}{(4s+1)(10s+1)}$
(c) Unstable $G_C(s) = \frac{s+1}{(2s-1)(4s+1)}$
    Let us now discretize them with a sampling period $T_0 = 2$ s

$$G_A(z) = \frac{0.0329z^{-1} + 0.0269z^{-2}}{1 - 1.4891z^{-1} + 0.5488z^{-2}} \qquad G_B(z) = \frac{-0.1017z^{-1} + 0.1730z^{-2}}{1 - 1.4253z^{-1} + 0.4966z^{-2}}$$

$$G_C(z) = \frac{-0.6624z^{-1} + 0.0137z^{-2}}{1 - 3.3248z^{-1} + 1.6487z^{-2}}$$

Depict step responses for the particular transfer functions and design appropriate PID–A1 type controllers. Verify dynamic behaviour of the closed loop systems by simulation for step changes of the reference signal $w(k)$ with $\xi = 1$ and $\omega_n = 0.45$.

(a) The model $G_A(z)$:
    The step response of the model $G_A(z)$ is shown in Figure 4.22. Using Table 4.3 and Equation (4.120) we can derive the controller equation

$$u(k) = 32.97e(k) - 39.14e(k-1) + 12.06e(k-2) + 0.409u(k-1) + 0.591u(k-2)$$

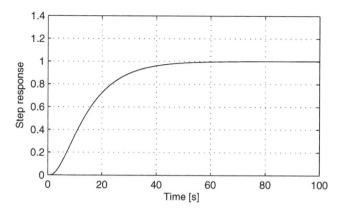

Figure 4.22. Step response of the model $G_A(z)$ – Example 4.1

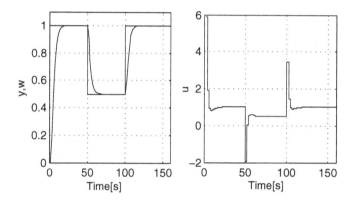

Figure 4.23. Example 4.1 – simulation results: control of the model $G_A(z)$

Simulated control responses of the model $G_A(z)$ are shown in Figure 4.23.

(b) The model $G_B(z)$:
The step response of the model $G_B(z)$ is shown in Figure 4.24. The controller output becomes

$$u(k) = 37.85e(k) - 48.58e(k-1) + 15.68e(k-2) - 4.461u(k-1) + 5.461u(k-2)$$

Simulated control responses of the model $G_B(z)$ are shown in Figure 4.25.

(c) The model $G_C(z)$:
The step response of the model $G_C(z)$ is shown in Figure 4.26. The controller output is given by the equation

$$u(k) = 5.27e(k) - 7.233e(k-1) + 2.484e(k-2) + 0.979u(k-1) + 0.021u(k-2)$$

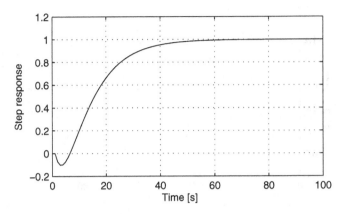

Figure 4.24. Example 4.1 – step response of the model $G_B(z)$

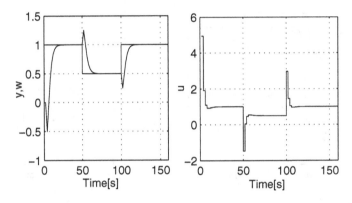

Figure 4.25. Example 4.1 – simulation results: control of the model $G_B(z)$

Simulated control responses of the model $G_C(z)$ are shown in Figure 4.27.

Example 4.2. Simulate control of the models used in the Example 4.1 using the PID–A1 type controller with recursive identification (self-tuning controller). The sampling period is set to $T_0 = 2$ s for all cases. Choose the initial parameter estimates as $\hat{\Theta}^T(0) = [0.1, 0.2, 0.1, 0.2]$ and the initial value of the directional forgetting factor as $\varphi(0) = 1$. Depict the evolution of the parameter estimates vector $\hat{\Theta}^T(k)$, the directional forgetting factor $\varphi(k)$ and the prediction error $\hat{e}(k)$. Simulate control of these models when using the convergent initial parameter estimates $\hat{\Theta}^T(k)$ in $k = 80$.

Make a simulation program with use of the Listings 3.1, 3.2 and Table 4.3. For computation of the controllers' parameters using the MATLAB® environment it is possible to apply the matrix Equation (4.127) and the expressions (4.126),(4.128). Control responses, together with the parameter estimates and

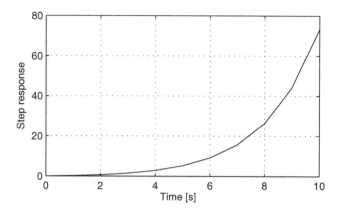

Figure 4.26. Example 4.1 – step response of the model $G_C(z)$

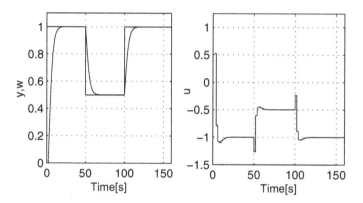

Figure 4.27. Example 4.1 – simulation results: control of the model $G_C(z)$

the evolution of the directional forgetting factor with the prediction error are shown in Figures 4.28–4.30.

Comparing these graphs we can clearly see the excellent effect of *a priori* information (the control process when using convergent estimates).

Structure of the PID–B Control Loop

The structure of this control loop with the block diagram illustrated in Figure 4.31 was designed by Ortega and Kelly [78]. Here, the controller equation takes the form

$$U(z) = [\beta E(z) - Q'(z^{-1})Y(z)]\frac{1}{P(z^{-1})} \qquad (4.133)$$

and while the polynomial $P(z^{-1})$ has the same form as polynomial (4.118) for controller (4.119), the second polynomial $Q'(z^{-1})$ takes the form

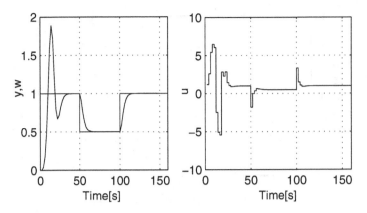

Figure 4.28. (a) Example 4.2 – control variables without *a priori* information: model $G_A(z)$

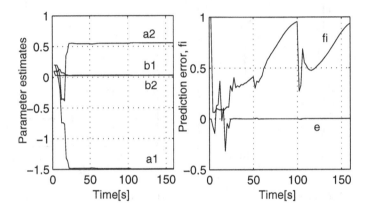

Figure 4.28. (b) Example 4.2 – parameter estimates, directional forgetting factor and prediction error: model $G_A(z)$

$$Q'(z^{-1}) = (1 - z^{-1})(q'_0 - q'_2 z^{-1}) \tag{4.134}$$

Substitution of polynomials (4.118) and (4.134) into Equation (4.133) yields the following relation for the controller output

$$u(k) = -[(q'_0 + \beta)y(k) - (q'_0 + q'_2)y(k - 1) + q'_2 y(k - 2)] - (\gamma - 1)u(k - 1)$$
$$+\gamma u(k - 2) + \beta w(k) \tag{4.135}$$

For the transfer function of the closed loop according to Figure 4.31 it is possible to obtain the relation

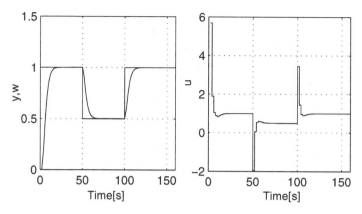

Figure 4.28. (c) Example 4.2 – control variables when using convergent estimates: model $G_A(z)$, initial parameter estimates $\hat{\Theta}^T(80) = [-1.4953, 0.5571, 0.0338, 0.0281]$

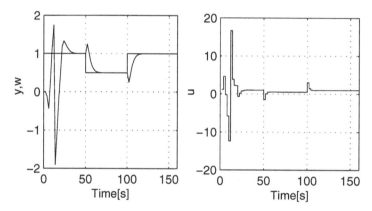

Figure 4.29. (a) Example 4.2 – control variables without *a priori* information: model $G_B(z)$

$$G_W(z) = \frac{Y(z)}{W(z)} = \frac{\beta B(z^{-1})}{A(z^{-1})P(z^{-1}) + B(z^{-1})[Q'(z^{-1}) + \beta]} \tag{4.136}$$

so the polynomial equation takes the form

$$A(z^{-1})P(z^{-1}) + B(z^{-1})[Q'(z^{-1}) + \beta] = D(z^{-1}) \tag{4.137}$$

In the case of controlled system polynomials in the form (4.116), polynomial Equation (4.137) defines a system of four linear algebraic equations with four unknown controller parameters q'_0, q'_2, β and γ:

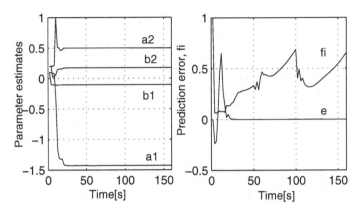

Figure 4.29. (b) Example 4.2 – Parameter estimates, directional forgetting factor and prediction error: model $G_B(z)$

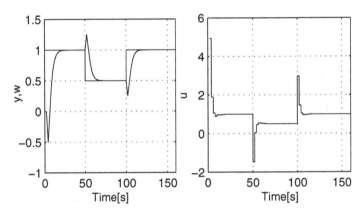

Figure 4.29. (c) Example 4.2 – control variables when using convergent estimates: model $G_B(z)$ Initial parameter estimates $\hat{\Theta}^T(80) = [-1.4283, 0.4996, -0.1018, 0.1732]$

$$
\begin{bmatrix}
b_1 & 0 & b_1 & 1 \\
b_2 - b_1 & -b_1 & b_2 & a_1 - 1 \\
b_2 & b_2 - b_1 & 0 & a_1 - a_2 \\
0 & b_2 & 0 & -a_2
\end{bmatrix}
\begin{bmatrix}
q_0' \\
q_2' \\
\beta \\
\gamma
\end{bmatrix}
=
\begin{bmatrix}
x_1 \\
x_2 \\
x_3 \\
x_4
\end{bmatrix}
\tag{4.138}
$$

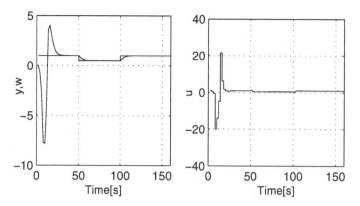

Figure 4.30. (a) Example 4.2 – control variables without *a priori* information: model $G_C(z)$

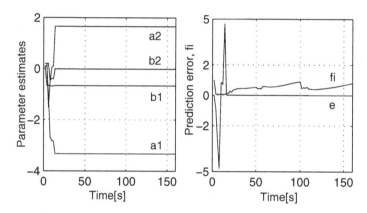

Figure 4.30. (b) Example 4.2 – parameter estimates, directional forgetting factor and prediction error: model $G_C(z)$

PID–B1 Controller

This type of controller is obtained when the polynomial $D(z^{-1})$ in polynomial Equation (4.137) is substituted by the relation (4.125) and the vector components on the right side of system Equation (4.138) are

$$x_1 = d_1 + 1 - a_1 \qquad x_2 = d_2 + a_1 - a_2 \qquad x_3 = -a_2 \qquad x_4 = 0 \quad (4.139)$$

By solving Equation system (4.138) we obtain the relations to calculate the controller parameters (see Table 4.4); variables d_1 and d_2 are calculated using relations (4.126).

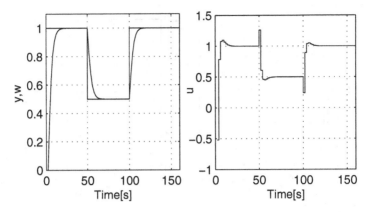

Figure 4.30. (c) Example 4.2 – control variables when using convergent estimates: model $G_C(z)$, initial parameter estimates $\hat{\Theta}^T(80) = [-3.3246, 1.6481, -0.6628, -0.0138]$

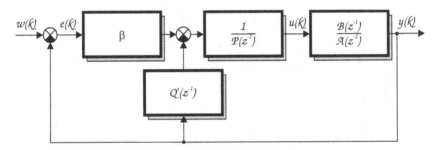

Figure 4.31. Block diagram of the control loop for a PID–B controller

Table 4.4. Relations to calculate PID–B controller parameters

Controller parameters	PID–B1	PID–B2
q_0'	$q_2'\left(\frac{b_1}{b_2} - \frac{a_1}{a_2}\right) - \frac{a_2}{b_2}$	$-\frac{r_2 - r_3 + r_4}{r_1}$
q_2'	$\frac{s_1}{r_1}$	$\frac{r_6 + r_7}{r_1}$
γ	$q_2'\frac{b_2}{a_2}$	$\frac{r_5}{r_1}$
β	$\frac{1}{b_1}(d_1 + 1 - a_1 - \gamma - b_1 q_0')$	$\frac{x_1 + x_2 - x_3 + x_4}{b_1 + b_2}$

PID–B2 Controller

The parameters for this type of a controller are obtained by inserting appropriately modified relation (4.129) as polynomial $D(z^{-1})$ into polynomial Equation (4.137), where the vector components on the right side of Equation system (4.138) is given by

$$x_1 = c + 1 - a_1 \qquad x_2 = d + a_1 - a_2 \qquad x_3 = -f - a_2 \qquad x_4 = g \quad (4.140)$$

and relations (4.131) are valid. In this case, by solving Equation system (4.138), we also obtain the equations for calculating the controller parameters given in Table 4.4 where the formulas to compute the auxiliary variables needed to determine the controller parameters take the form

$$
\begin{aligned}
s_1 &= a_2\{b_2[a_1(b_1 + b_2) + b_1(d_2 - a_2) - b_2(d_1 + 1)] - a_2 b_1^2\} \\
r_1 &= (b_1 + b_2)(a_1 b_1 b_2 - a_2 b_1^2 - b_2^2) \\
r_2 &= a_1 b_2[b_1(x_2 - x_3 + x_4) - b_2 x_1] \\
r_3 &= a_2 b_1[b_2 x_1 - b_1(x_2 - x_3 + x_4)] \\
r_4 &= (b_1 + b_2)[b_1 x_4 + b_2(x_3 - x_4)] \\
r_5 &= b_1(b_1^2 x_4 + b_1 b_2 x_3 + b_2^2 x_2) - b_2^3 x_1 \\
r_6 &= b_1^2(a_2 x_3 + a_1 x_4 - a_2 x_4) \\
r_7 &= b_2[b_1(a_1 x_4 + a_2 x_2 - x_4) - b_2(a_2 x_1 + x_4)]
\end{aligned}
\qquad (4.141)
$$

4.7.4 Digital PID Controllers Based on the Modified Ziegler–Nichols Criterion

The experimental tuning of parameters for a continuous-time PID controller designed by Ziegler and Nichols [103] more than half a century ago is still used today in industrial practice. In this well-known and popular approach the PID controller parameters are calculated from the critical (ultimate) proportional gain K_{Pu} and the critical (ultimate) period of oscillations T_u of a closed loop. These critical parameters are obtained by gradually increasing the gain of the proportional controller until the output of the closed loop oscillates at constant amplitude, *i.e.* the control loop is at the stability boundary. In this case the poles of the closed loop are placed on the imaginary axis of a complex s-plane. Then both, the proportional critical gain K_{Pu} and the critical period of oscillations T_u are recorded. The parameters of the PID controller are determined by the relations

$$K_P = 0.6 K_{Pu} \qquad T_I = 0.5 T_u \qquad T_D = 0.125 T_u \qquad (4.142)$$

The following relations are recommended to calculate the parameters for PID controller (3.38) [56]

$$K_P = 0.6 K_{Pu}\left(1 - \frac{T_0}{T_u}\right) \qquad T_I = \frac{K_P T_u}{1.2 K_{Pu}} \qquad T_D = \frac{3 K_{Pu} T_u}{40 K_P} \qquad (4.143)$$

The disadvantage of experimentally determining the critical parameters is that the system can be brought to a state of instability, and the finding stability boundary in systems with large time constants can be very time-consuming. The modified method given below for tuning a digital PID controller avoids this problem.

When discretizing a control loop, the continuous-time controller output is modified to the form of a step function using a sample and hold unit. Then the step function can approximate the original continuous-time signal delayed by half of a sampling period T_0. More simply, we can assume that the system's discrete model differs from the continuous-time model in that it includes an extra time delay of $T_0/2$. The time delay does not change amplitude but linearly increases the phase shift as the frequency grows.

$$\varphi = -\frac{T_0\omega}{2} \tag{4.144}$$

At critical frequency ω_u the system has phase shift $\varphi = -\pi$ and gain A_u, validating

$$A_u K_{Pu} = -1 \tag{4.145}$$

In discrete control the effects of phase shift φ caused by discretization changes the critical frequency and, since the system has different gain at different frequencies, critical gain will also change. The critical values depend on the sampling period chosen and this will be referred to as function T_0, $i.e.$ $K_{Pu}(T_0)$ and $T_u(T_0)$.

Calculating the Critical Parameters for an n-th order Model

Assume the discrete transfer function of the controlled process taking the form

$$G_P(z) = \frac{Y(z)}{U(z)} = \frac{z^{-d}B(z^{-1})}{A(z^{-1})} \tag{4.146}$$

with polynomials

$$A(z^{-1}) = 1 + \sum_{i=1}^{n} a_i z^{-i} = 1 + a_1 z^{-1} + a_2 z^{-2} + \ldots + a_n z^{-n} \tag{4.147}$$

$$B(z^{-1}) = \sum_{i=1}^{n} b_i z^{-i} = b_1 z^{-1} + b_2 z^{-2} + \ldots + b_n z^{-n} \tag{4.148}$$

where d is the number of time delay intervals. Next, consider the discrete transfer function of a proportional controller

$$G_R(z) = \frac{U(z)}{E(z)} = K_P \tag{4.149}$$

The transfer function of the closed loop, illustrated as a block diagram in Figure 4.20, then takes the form

$$G_w(z) = \frac{Y(z)}{W(z)} = \frac{G_P(z)G_R(z)}{1 + G_P(z)G_R(z)} = \frac{z^{-d}K_pB(z^{-1})}{A(z^{-1}) + z^{-d}K_pB(z^{-1})} \quad (4.150)$$

The denominator of transfer function (4.150) is the characteristic polynomial

$$D(z^{-1}) = A(z^{-1}) + z^{-d}K_pB(z^{-1}) \quad (4.151)$$

the poles of which determine the dynamic behaviour of the closed loop. It is clear that the closed loop will be at the stability boundary if at least one of the poles of the characteristic polynomial (4.151) is placed on the unit circle and the others are inside it. This meets condition $K_P = K_{Pu}(T_0)$. There are two possible positions for pole assignment of the characteristic polynomial on the unit circle (Figure 4.32) may occur so that the control loop is on the stability boundary.

- The characteristic polynomial (4.151) has a pair of complex conjugate poles $z_{1,2} = \alpha \pm j\beta$, for which $\alpha^2 + \beta^2 = 1$. Thus, the appropriate part of the characteristic polynomial can be expressed as a product of root factors

$$D_1(z) = (z - z_1)(z - z_2) = z^2 - 2\alpha z + 1 \quad (4.152)$$

- The characteristic polynomial (4.151) has one or more real poles $z_j = \alpha_j = -1; (\beta_j = 0)$ so that the appropriate part of the characteristic polynomial can be expressed in the form

$$D_2(z) = (z + 1)^j \quad (4.153)$$

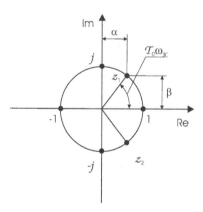

Figure 4.32. Placement of critical poles on the unit circle

In the first case, characteristic polynomial (4.151) must be divisible by polynomial (4.152), which leads to the solution of the polynomial equation

$$z^{n+d}\left[A(z^{-1}) + z^{-d}K_{Pu}(T_0)B(z^{-1})\right] = (z^2 - 2\alpha z + 1)z^{n+d-2}E(z^{-1}) \quad (4.154)$$

where

$$E(z^{-1}) = 1 + \sum_{i=1}^{n+d-2} e_i z^{-i} \tag{4.155}$$

and $K_{Pu}(T_0)$, α and e_i are the unknown parameters of polynomial Equation (4.154).

The following polynomial equation can be obtained for the second case

$$z^{n+d} \left[A(z^{-1}) + z^{-d} K_{Pu}(T_0) B(z^{-1}) \right] = (z+1)^j z^{n+d-j} F(z^{-1}) \tag{4.156}$$

where

$$F(z^{-1}) = 1 + \sum_{i=1}^{n+d-j} f_i z^{-i} \tag{4.157}$$

and $K_{Pu}(T_0)$, f_i are the unknown parameters of polynomial Equation (4.156). In both cases the solution with $K_{Pu}(T_0) > 0$ is chosen, which does not contain unstable poles.

Rather than solve Equation (4.146) it is possible to calculate critical gain $K_{Pu}(T_0)$ in a more simple way. For the pole $\alpha = -1$ the oscillating component is created by the term $(-1)^k$, which corresponds to the continuous-time function $\cos \frac{\pi}{T_0} t$. Here, the critical frequency is given by the relation

$$\omega_u = \frac{\pi}{T_0} \tag{4.158}$$

If we insert critical frequency ω_u (according to Equation (4.158)) into the defining relation of the z-transform

$$z = e^{j\omega T_0} = \cos \omega T_0 + j \sin \omega T_0 \tag{4.159}$$

then the solution to Equation (4.159) is $z = -1$. As a result, critical gain is calculated using equation

$$K_{Pu}(T_0) = -\frac{1}{G_P(z^{-1})} = -\frac{1}{G_P(-1)} \tag{4.160}$$

Solving equations (4.154) and (4.156) using the method of uncertain coefficients for $n + d > 3$ involves fairly complicated calculations so it is advisable to use a numeric method to solve the polynomial equation.

Again, calculation of the critical period of oscillations depends on placement of the poles on the unit circle in the complex z-plane. It is clear from Figure 4.32 that the critical period of oscillations can be computed from the relations

$$\cos(T_0 \omega_u) = \alpha \qquad \omega_u = \frac{1}{T_0} \arccos \alpha \qquad T_u(T_0) = \frac{2\pi}{\omega_u} \tag{4.161}$$

Obviously, for real critical poles $z_j = -1$, the following formulas for the critical period of oscillations are valid

$$\cos(T_0\omega_u) = -1 \qquad \omega_u = \frac{\pi}{T_0} \qquad T_u(T_0) = 2T_0 \qquad (4.162)$$

Using the unified approach given above it is also possible to derive algorithms for control systems with time delay. This of course holds only for lower-order systems with few time delay steps, since with each increase in the order of the system or number of time delay steps there is an increase in the degree of the characteristic polynomial (4.151).

Now, let us derive the relations to calculate the critical proportional gain for the first- to third-order models. It will be shown that for second-order models it is possible to compute the critical parameters using other approaches.

Calculating Critical Gain for First-order Models

Let the process be described by a first-order model ($n = 1, d = 0$ in Equations (4.146)–(4.148)). Characteristic polynomial (4.151) then takes the form

$$D(z) = z + a_1 + K_P b_1 \qquad (4.163)$$

When using a first-order model only one critical real pole can exist and so polynomial Equation (4.156) becomes

$$z + a_1 + K_{Pu}(T_0)b_1 = z + 1 \qquad (4.164)$$

where the critical gain is given by the relation

$$K_{Pu}(T_0) = \frac{1 - a_1}{b_1} \qquad (4.165)$$

Calculating Critical Gain for Second-order Models

There are several methods for deriving the relations to calculate the critical gain.

Unified Approach
Let the process be described by a second-order model ($n = 2, d = 0$ in Equations (4.146)–(4.148)). Characteristic polynomial (4.151) then takes the form

$$D(z) = z^2 + (a_1 + b_1 K_P)z + (a_2 + b_2 K_P) \qquad (4.166)$$

In the first case, *i.e.* when considering polynomial Equation (4.154), this equation, as applied to a second-order model, will take the form

$$z^2\left\{1 + a_1 z^{-1} + a_2 z^{-2} + K_{Pu}(T_0)[b_1 z^{-1} + b_2 z^{-2}]\right\} = z^2 - 2\alpha z + 1 \quad (4.167)$$

By comparing the coefficients of the same power of z, it is possible to obtain the two equations

$$a_1 + K_{Pu}(T_0)b_1 = -2\alpha \qquad a_2 + K_{Pu}(T_0)b_2 = 1 \qquad (4.168)$$

Then, the relations to calculate critical gain and real parts of the complex conjugate pole are obtained from the equations above and have the form

$$K_{Pu}(T_0) = \frac{1 - a_2}{b_2} \qquad \alpha = \frac{a_2 b_1 - a_1 b_2 - b_1}{2b_2} \qquad (4.169)$$

In the second case, *i.e.* when considering polynomial Equation (4.156), the equation for a second-order model takes the form

$$z^2\{1 + a_1 z^{-1} + a_2 z^{-2} + K_{Pu}(T_0)[b_1 z^{-1} + b_2 z^{-2}]\} = (z+1)z(1 + f_1 z^{-1}) \quad (4.170)$$

In the same way as in the previous example we obtain two equations

$$a_1 + K_{Pu}(T_0)b_1 = 1 + f_1 \qquad a_2 + K_{Pu}(T_0)b_2 = f_1 \qquad (4.171)$$

defining the critical gain

$$K_{Pu}(T_0) = \frac{a_1 - a_2 - 1}{b_2 - b_1} \qquad (4.172)$$

The relation can also be derived from Equation (4.160).

The critical gain is calculated either from Equation (4.169) or (4.172) depending on the condition

$$b^2 - 4c < 0 \qquad (4.173)$$

where expression (4.173) is the discriminant of the characteristic equation

$$z^2 + (a_1 + b_1 K_P)z + (a_2 + b_2 K_P) = 0 \qquad (4.174)$$

with notation

$$b = a_1 + b_1 K_P \qquad c = a_2 + b_2 K_P \qquad (4.175)$$

A flow diagram for calculation of the controller parameters for a second-order system is illustrated in Figure 4.33. The MATLAB® code of this algorithm is given in Listing 4.1.

Complex Analysis of the Placement of Critical Poles in the z-Plane
If we denote characteristic polynomial

$$D(z) = z^2 + bz + c \qquad (4.176)$$

with coefficients as in (4.175), there are four possibilities for pole assignment of the second-order characteristic polynomial so that the closed loop is on the stability boundary (Figure 4.34):

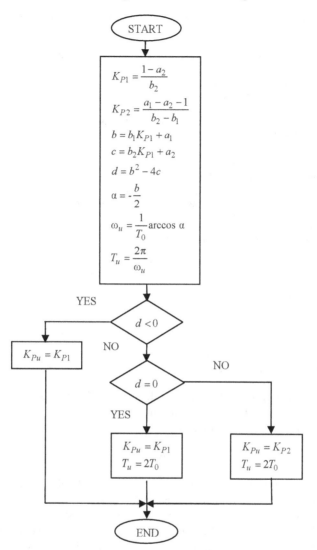

Figure 4.33. Flow diagram for the calculation of critical parameters for a second-order system

(a) The characteristic polynomial (4.176) has a pair of complex conjugate poles $z_{1,2} = \alpha \pm j\beta$ (Figure 4.34a)) where the formula $\alpha^2 + \beta^2 = 1$ holds. Then, the characteristic polynomial (4.176) can be expressed as a product of root factors

$$D(z) = (z - z_1)(z - z_2) \qquad (4.177)$$

Substitution of $z_{1,2} = \alpha \pm j\beta$ and $\alpha^2 + \beta^2 = 1$ into Equation (4.177) yields

Listing 4.1. Calculation of critical parameters for a second-order system

```
%Calculation of critical parameters for a 2nd-order system
Kp1=(1-a2)/b2; Kp2=(a1-a2-1)/(b2-b1);
bb=b1*Kp1+a1; cc=b2*Kp1+a2;
dd=bb*bb-4*cc; alfa=-bb/2;
if alfa>1
    omegau=(1/T0)*acos(.99);  %critical frequency
elseif alfa<-1
    omegau=(1/T0)*acos(-1);
else
    omegau=(1/T0)*acos(alfa);
end
Tu=(2*pi)/omegau; %critical period of oscillations

if dd<0
        Kpu=Kp1; %critical gain
    elseif dd==0
        Kpu=Kp1; Tu=2*T0;
    else
        Kpu=Kp2;    Tu=2*T0;
    end
```

$$D(z) = (z-\alpha-j\beta)(z-\alpha+j\beta) = z^2 - 2\alpha z + \alpha^2 + \beta^2 = z^2 - 2\alpha z + 1 \quad (4.178)$$

Comparing Equations (4.176) and (4.178) it follows that $b = -2\alpha$ (therefore $\alpha = -b/2$) and $c = 1$. Assuming $K_P = K_{Pu}(T_0)$ and replacing the variable c from Equation (4.175) with

$$a_2 + b_2 K_{Pu}(T_0) = 1 \quad (4.179)$$

it is possible to obtain a relation to calculate critical gain and a formula for the real part of complex conjugate pole (4.169).

(b) The characteristic polynomial (4.176) has a double real pole $z_{3,4} = \alpha$ (imaginary part $\beta = 0$) – Figure 4.34b. The control loop is on the stability boundary only in the case when $\alpha = -1$, because the positive real pole $\alpha = 1$ will not cause the control loop to oscillate. As a result, the characteristic polynomial (4.176), expressed as a product of root factors, now takes the form

$$D(z) = (z+1)^2 = z^2 + 2z + 1 \quad (4.180)$$

Since $c = 1$, therefore relation (4.169) can again be used for computing critical gain $K_{Pu}(T_0)$.

(c) The characteristic polynomial (4.176) has only imaginary poles $z_{5,6} = \pm j$ (real component $\alpha = 0$ – Figure 4.34c) so the characteristic polynomial (4.176) takes the form

$$D(z) = (z+j)(z-j) = z^2 + 1 \quad (4.181)$$

It is clear from Equation (4.181) that $b = 0$ and $c = 1$, therefore the first relation for the calculation of K_{Pu} (4.169) is also valid.

(d) Characteristic polynomial (4.176) has one pole $z_7 = -1$ and a second real pole $|z_8| < 1$ (a stable pole inside the unit circle – Figure 4.34d). The characteristic polynomial (4.176) can again be expressed as a product of root factors

$$D(z) = (z + 1)(z - z_8) \qquad (4.182)$$

It follows from Equations (4.176) and (4.182) that the characteristic polynomial (4.176) should be divisible by the factor $z + 1$ without any remainder. Since the following relation

$$\frac{z^2 + bz + c}{z + 1} = z + b - 1 + \frac{1 - b + c}{z + 1} \qquad (4.183)$$

holds, the condition of zero remainder is fulfilled in the case

$$1 - b + c = 0 \qquad (4.184)$$

Upon substituting from Equation (4.175) for the variables b and c and $K_P = K_{Pu}(T_0)$ into Equation (4.184), it yields

$$1 - a_1 - K_{Pu}(T_0)b_1 + a_2 + K_{Pu}(T_0)b_2 = 0 \qquad (4.185)$$

and after some manipulations, relation (4.172) is obtained. The pole $z_8 = 1 - b$ lies inside the unit circle.

The critical gain $K_{Pu}(T_0)$ is then computed either from the first relation of Equations (4.169) or from Equation (4.172) according to fulfilment of the condition (4.173). It is obvious that when satisfying the condition (4.173), the characteristic polynomial (4.176) has its poles situated according to (a), (b) or (c). Therefore, for the computation of $K_{Pu}(T_0)$ the first relation of equations (4.169) is used. If condition (4.173) is not fulfilled, the characteristic polynomial (4.176) has its poles situated according to (d) and $K_{Pu}(T_0)$ is computed according to relation (4.172).

Derivation Using Bilinear Transformation

It is possible to achieve the same relations using bilinear transformation

$$z = \frac{w + 1}{w - 1} \qquad (w = \alpha + j\beta) \qquad (4.186)$$

where the unit circle in the z-plane transforms into the imaginary axis of the w-plane so that point $(-1; 0)$ in the z-plane corresponds to point $(0; 0)$ of the w-plane. The introduction of transformation (4.186) into characteristic Equation (4.174) yields quadratic equation

$$w^2(1 + b + c) + w(2 - 2c) + 1 - b + c = 0 \qquad (4.187)$$

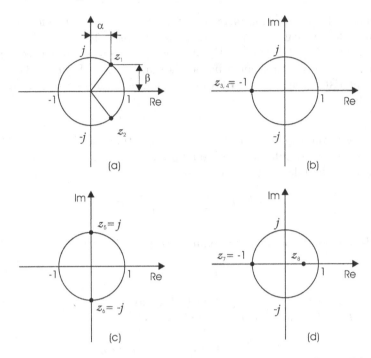

Figure 4.34. Placement of critical poles for a second-order model

With regard to stability, Equation (4.187) can be solved using a well-known method from the theory of continuous-time systems. A continuous-time control loop is at the stability boundary if roots of characteristic Equation (4.187) lie on the imaginary axis of the w complex plane, *i.e.* when $\alpha = 0$. Then Equation (4.187) takes a form

$$-\beta^2(1 + b + c) + j\beta(2 - 2c) + 1 - b + c = 0 \tag{4.188}$$

and can be separated into real and imaginary parts

$$-\beta^2(1 + b + c) + 1 - b + c = 0 \tag{4.189a}$$
$$\beta(1 - c) = 0 \tag{4.189b}$$

Equation (4.189) has two solutions:

- $1 - c = 0$, from which it follows that $c = 1$ and by inserting this into the second Equation of (4.175) it is possible to obtain the first relation of Equation (4.169) to calculate the critical gain;
- $\beta = 0$, which can be put into Equation (4.189a) to obtain formula (4.184) so that the relation to calculate the critical gain (4.172) holds.

A demonstration of the algorithm for this kind of a self-tuning controller is presented here. The ARX model in the form (3.10) is used in the identification part of the control algorithm, where

$$\hat{\boldsymbol{\Theta}}^T(k) = \left[\hat{a}_1, \hat{a}_2, \hat{b}_1, \hat{b}_2\right] \tag{4.190}$$

is the parameter estimates vector and

$$\phi^T(k-1) = [-y(k-1), -y(k-2), u(k-1), u(k-2)] \tag{4.191}$$

is the regression vector. As an example, the Takahashi controller (4.38) is used, which can be modified to the form below

$$u(k) = K_R[y(k-1) - y(k)] + K_I[w(k) - y(k)]$$
$$+K_D[2y(k-1) - y(k-2) - y(k)] + u(k-1) \tag{4.192}$$

where

$$K_R = 0.6K_{Pu} - \frac{K_I}{2} \qquad K_I = \frac{1.2K_{Pu}T_0}{T_u} \qquad K_D = \frac{3K_{Pu}T_u}{40T_0} \tag{4.193}$$

are the adjustable components of a digital PID controller taken from [56]. If we introduce the following notation for the individual elements of regression vector (4.191)

$$d_1 = y(k-1), \qquad d_2 = y(k-2), \qquad d_3 = u(k-1), \qquad d_4 = u(k-2)$$

then Equation (4.192) can be rewritten into the form

$$u(k) = K_R[d_1 - y(k)] + K_I[w(k) - y(k)] + K_D[2d_1 - d_2 - y(k)] + d_3 \tag{4.194}$$

It is clear from the first relation of Equations (4.169) that division by zero will occur when $\hat{b}_2 = 0$ and this will also happen in relation (4.172) for $\hat{b}_1 = \hat{b}_2$. It is also necessary to take into account the fact that the second relation of Equation (4.161) must meet condition

$$\alpha = \left|-\frac{b}{2}\right| \leq 1 \tag{4.195}$$

The algorithm of this controller now consists of the following steps performed at each sampling period:

Step 1. Parameter estimates of the process model (4.190).

Step 2. If parameter estimate $\hat{b}_2$ is less than machine zero or if $\hat{b}_1 = \hat{b}_2$, use the previous parameter estimate $\hat{b}_2$.

Step 3. Calculation of critical gain $K_{Pu}(T_0)$ and critical period $T_u(T_0)$ (see the flow diagram, Figure 4.30). If $\alpha < -1$ set $\alpha = -1$; if $\alpha > 1$, set $\alpha = 1$ (see relation (4.195)).

Step 4. Calculation of digital PID controller parameters according to relations (4.193).

Step 5. Calculation of controller output according to relation (4.194).
Step 6. Limiting controller output $u(k)$ with respect to the actuator constraints and technological conditions.
Step 7. Cyclic exchange of the data in regression vector (4.191)

$$d_4 = d_3, \qquad d_2 = d_1, \qquad d_3 = u(k), \qquad d_1 = -y(k)$$

Example 4.3. For the models used in the Example 4.1 design controllers based on the modified Ziegler–Nichols method. Verify by simulation the dynamic behaviour of the closed loops using the Takahashi controller (4.192), (4.193) for step changes of the reference signal $w(k)$.

Using the diagram in Figure 4.33 or the Listing 4.1 it is possible to calculate appropriate critical parameters $K_{P1}(2)$ and $K_{P2}(2)$. Consequently we can calculate the critical gain $K_u(2)$ and the critical period of oscillation $T_u(2)$.

(a) The model $G_A(z)$:
   The critical parameters are: $K_{P1}(2) = 16.7732$; $K_{P2}(2) = 506.3172$; $K_u(2) = 16.7732$; $T_u(2) = 11.6027$. The controller output is given by the equation

$$u(k) = 8.3292[y(k-1) - y(k)] + 3.4695[w(k) - y(k)]$$
$$+7.2980[2y(k-1) - y(k-2) - y(k)] + u(k-1)$$

Simulation results of the control of the model $G_A(z)$ are shown in Figure 4.35.

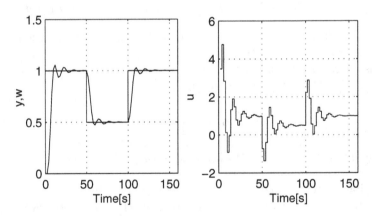

Figure 4.35. Example 4.3 – simulation results: control of the model $G_A(z)$

(b) The model $G_B(z)$:
   The critical parameters are: $K_{P1}(2) = 2.2098$; $K_{P2}(2) = -10.6367$; $K_u(2) = 2.9098$; $T_u(2) = 23.5184$. The controller output is given by

$$u(k) = 1.5974[y(k-1) - y(k)] + 0.2969[w(k) - y(k)]$$
$$+ 2.5663[2y(k-1) - y(k-2) - y(k)] + u(k-1)$$

Simulation results for the model $G_B(z)$ are shown in Figure 4.36.

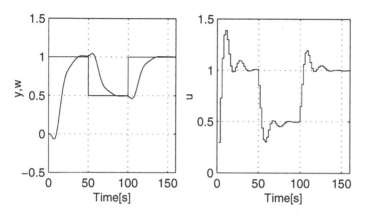

Figure 4.36. Example 4.3 – simulation results: control of the model $G_B(z)$

(c) The model $G_C(z)$:
Control of the model $G_C(z)$ is unstable.

Example 4.4. Simulate control of the models $G_A(z)$ and $G_B(z)$ used in the Example 4.1 employing a self-tuning controller based on the modified Ziegler–Nichols method. Verify by simulation the dynamic behaviour of the closed loops using the Takahashi controller (4.192), (4.193). The sampling period for both cases is $T_0 = 2$ s. Choose the initial estimates of model parameters as $\hat{\Theta}^T(0) = [0.1, 0.2, 0.1, 0.2]$ and the initial value of the directional forgetting factor as $\varphi(0) = 1$. Simulate also control of these models when using the convergent initial parameter estimates $\hat{\Theta}^T(k)$ in $k = 80$. Design a simulation program using the Listings 3.1, 3.2 and 4.1. Simulation results for models $G_A(z)$ and $G_B(z)$ are shown in Figures 4.37 and 4.38.

The MATLAB® program for control of the model $G_A(z)$ is given in Listing 4.2.

Calculating Critical Gain for Third-order Models

The relations needed to calculate critical gain in a process described by a third-order model ($n = 3, d = 0$ in Equations (4.146)–(4.148)) can be derived similarly to those in previous sections. Here, only the unified approach is

Listing 4.2. Self-tuning Takahashi controller for a second-order model

```
%Simulation verification of self-tuning Takahashi controller
%G(s)=1/(5s+1)(10s+1)      %Controlled continuous-time model
T0=2;                      %Sampling period
A1=-1.4891; A2=0.5488;
B1=0.0329; B2=0.0269;
[d,theta,c,ro,fi,la,ny,u]=inidedf;      %Initialization
D=zeros(4,1);                %[y(k-1), y(k-2), u(k-1), u(k-2)]
w=1;                         % Choice of reference signal
for k=1:80
   if k==25     w=0.5;
   elseif k==50     w=1;
   end
   steps(k)=k;
   y(k)=-A1*D(1)-A2*D(2)+B1*D(3)+B2*D(4);     %Simulation

   % Recursive identification
   [fi,theta,d,c,la,ny,ep,te,ks,pp]
   =identdf(fi,theta,d,c,la,ny,D(3),y(k),ro);

   % Parameter estimates
   a1=theta(1); a2=theta(2); b1=theta(3); b2=theta(4);

   Kp1=(1-a2)/b2;  Kp2=(a1-a2-1)/(b2-b1);  %critical param. computation
   bb=b1*Kp1+a1;   cc=b2*Kp1+a2;   dd=bb*bb-4*cc;   alfa=-bb/2;

   if alfa>1    omegau=(1/T0)*acos(.99);
   elseif alfa<-1    omegau=(1/T0)*acos(-1);
   else    omegau=(1/T0)*acos(alfa);
   end
   Tu=(2*pi)/omegau;

   if dd<0    Kpu=Kp1; %critical gain
   elseif dd==0    Kpu=Kp1;    Tu=2*T0;
   else    Kpu=Kp2;    Tu=2*T0;
   end

   kp=0.6*Kpu*(1-T0/Tu);                    %controller
   ki=1.2*Kpu*T0/Tu;
   kd=3*Kpu*Tu/(40*T0);

   u(k)=kp*(D(1)-y(k))+ki*(w-y(k))+kd*(2*D(1)-D(2)-y(k))+D(3);

   % Cyclic date substitution in process model
   D(2)=D(1);  D(1)=y(k);  D(4)=D(3);  D(3)=u(k);

   ww(k)=w;                          %save control loop variables
   aa1(k)=a1; aa2(k)=a2; bb1(k)=b1; bb2(k)=b2;
   ff(k)=fi; eep(k)=ep;
end;
plot(steps,y,'m', steps,u,'r', steps,ww,'g');
figure;
plot(steps,aa1,'m',steps,aa2,'r',steps,bb1,'g',steps,bb2,'b');
```

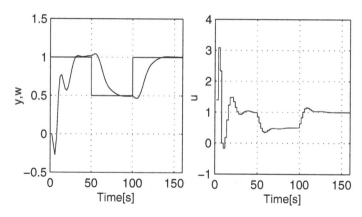

Figure 4.37. (a) Example 4.4 – control variables without *a priori* information: model $G_A(z)$

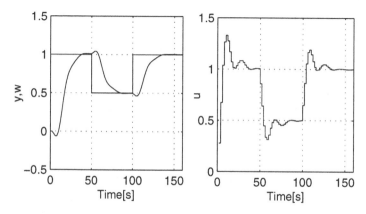

Figure 4.37. (b) Example 4.4 – control variables when using convergent estimates: model $G_A(z)$, initial parameter estimates $\hat{\Theta}^T(80) = [-1.5060, 0.5673, 0.0307, 0.0307]$

employed, although in this case it is also possible to use bilinear transformation methods. Characteristic polynomial (4.151) has the form

$$D(z) = z^3 + (a_1 + b_1 K_P)z^2 + (a_2 + b_2 K_P)z + a_3 + b_3 K_P \qquad (4.196)$$

In the first case, *i.e.* when considering polynomial Equation (4.154), the equation for $K_P = K_{Pu}(T_0)$, when applied to a third-order model, takes the form

$$z^3\{1 + a_1 z^{-1} + a_2 z^{-2} + a_3 z^{-3} + K_{Pu}(T_0)[b_1 z^{-1} + b_2 z^{-2} + b_3 z^{-3}]\}$$
$$= (z^2 - 2\alpha z + 1)z(1 + e_1 z^{-1}) \qquad (4.197)$$

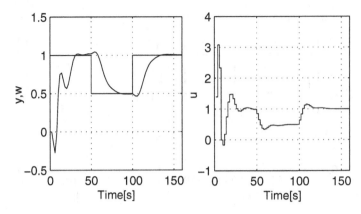

Figure 4.38. (a) Example 4.4 – control variables without *a priori* information: model $G_B(z)$

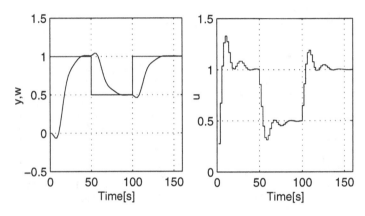

Figure 4.38. (b) Example 4.4 – Control variables when using convergent estimates: model $G_B(z)$, initial parameter estimates $\hat{\Theta}^T(80) = [-1.4345, 0.5118, -0.1043, 0.1817]$

By comparing coefficients at the same power of z, it is possible to obtain three equations with three unknowns $K_{Pu}(T_0), \alpha$ and e_1

$$a_1 + b_1 K_{Pu}(T_0) = -2\alpha + e_1$$
$$a_2 + b_2 K_{Pu}(T_0) = 1 - 2\alpha e_1 \qquad (4.198)$$
$$a_3 + b_3 K_{Pu}(T_0) = e_1$$

Equation (4.198) has the following solution

$$K_{P1,2}(T_0) = \frac{-r_1 \pm \sqrt{d}}{2r_2} \qquad \alpha = \frac{a_3 - a_1 + K_{Pu}(T_0)[b_3 - b_1]}{2} \qquad (4.199)$$

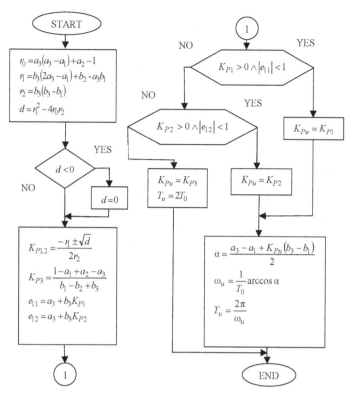

Figure 4.39. Flow diagram of the calculation of critical parameters for a third-order model

where

$$r_0 = a_3(a_3 - a_1) + a_2 - 1 \qquad r_1 = b_3(2a_3 - a_1) + b_2 - a_3b_1;$$
$$r_2 = b_3(b_3 - b_1) \qquad d = r_1^2 - 4r_0r_2 \tag{4.200}$$

and the last relation of Equation (4.198) is used to calculate e_1. Then, the solution which satisfies following inequalities

$$K_{Pu}(T_0) > 0 \qquad |e_1| < 1 \tag{4.201}$$

is taken. If conditions (4.201) are not met either for $K_{P1}(T_0)$ or $K_{P2}(T_0)$, Equation (4.156) must be used to calculate the critical gain. For a third-order model this takes the form

$$z^3\{1 + a_1z^{-1} + a_2z^{-2} + a_3z^{-3} + K_{Pu}(T_0)[b_1z^{-1} + b_2z^{-2} + b_3z^{-3}]\}$$
$$= (z+1)z^2(1 + f_1z^{-1} + f_2z^{-2}) \tag{4.202}$$

Comparing coefficients at the same power of z yields the following three equations

$$a_1 + b_1 K_{Pu}(T_0) = 1 + f_1$$
$$a_2 + b_2 K_{Pu}(T_0) = f_1 + f_2 \qquad (4.203)$$
$$a_3 + b_3 K_{Pu}(T_0) = f_2$$

and by solving these, a relation for critical gain is obtained as follows

$$K_{Pu}(T_0) = \frac{1 - a_1 + a_2 - a_3}{b_1 - b_2 + b_3} \qquad (4.204)$$

The relation (4.204) can also be obtained by solving Equation (4.160). A flow diagram showing the calculation of controller parameters is given in Figure 4.39, and MATLAB® code is given in Listing 4.3.

The algorithm of this self-tuning controller has a similar structure to that of a second-order model. Again, the ARX model in form (3.10) is used in the identification part, where the vector of parameter estimates for the process model has the form

$$\hat{\Theta}^T(k) = \begin{bmatrix} \hat{a}_1, \hat{a}_2, \hat{a}_3, \hat{b}_1, \hat{b}_2, \hat{b}_3 \end{bmatrix} \qquad (4.205)$$

and

$$\phi^T(k-1) = [-y(k-1), -y(k-2), -y(k-3), u(k-1), u(k-2), u(k-3)] \qquad (4.206)$$

is the regressor.

Example 4.5. Consider a continuous-time transfer function as a model of a controlled process

$$G_D(s) = \frac{1}{(0.5s + 1)(s + 1)(2s + 1)}$$

Then its discrete version for the sampling period $T_0 = 1$ s has the form

$$G_D(z) = \frac{0.0706z^{-1} + 0.1416z^{-2} + 0.0136z^{-3}}{1 - 1.2747z^{-1} + 0.5361z^{-2} + 0.0302z^{-3}}$$

Design a controller based on the modified Ziegler–Nichols method. Verify by simulation dynamic behaviour of the closed loop using Takahashi controller (4.192), (4.193) for step changes of the reference signal $w(k)$. Using the flow diagram in Figure 4.37 or the Listing 4.3 we can calculate appropriate critical parameters $K_{P1}(1), K_{P2}(1)$ and $K_{P3}(1)$. Consequently, we can compute the critical gain $K_u(1)$ and the critical period of oscillations $T_u(1)$. Then, the critical parameters have these values: $K_{P1}(1) = 3.1782; K_{P2}(1) = 203.5381; K_{P3}(1) = -49.4948; K_u(1) = 3.1783; T_u(1) = 11.6027$ and the controller output is computed from equation

Listing 4.3. Calculation of critical parameters for a third-order system

```
%Calculation of critical gain Kpu
r2=b3*(b3-b1); r1=b3*(2*a3-a1)+b2-a3*b1; r0=a3*(a3-a1)+a2-1;
dd=r1*r1-4*r0*r2;
if dd<0    dd=0;
end
Kp1=(-r1+sqrt(dd))/(2*r2);
Kp2=(-r1-sqrt(dd))/(2*r2);
Kp3=(a1+a3-a2-1)/(b2-b1-b3);
Kpu=Kp1;
aa=b1*Kpu+a1;   bb=b2*Kpu+a2;  cc=b3*Kpu+a3;
pp=bb-aa*aa/3;  qq=(2*aa*aa*aa/27)-(aa*bb/3)+cc;
qq1=(qq/2)*(qq/2);  pp1=(pp/3)*(pp/3)*(pp/3);
DD=qq1+pp1;
if DD<=0
    Kpu=Kp2;
    aa=b1*Kpu+a1;  bb=b2*Kpu+a2;  cc=b3*Kpu+a3;
    pp=bb-aa*aa/3;  qq=2*aa*aa*aa/27-aa*bb/3+cc;
    qq1=(qq/2)*(qq/2);  pp1=(pp/3)*(pp/3)*(pp/3);
    DD=qq1+pp1;
end
if DD<=0
    Kpu=Kp3;
    aa=b1*Kpu+a1;  bb=b2*Kpu+a2;  cc=b3*Kpu+a3;
    pp=bb-aa*aa/3;  qq=2*aa*aa*aa/27-(aa*bb/3)+cc;
    qq1=(qq/2)*(qq/2);  pp1=(pp/3)*(pp/3)*(pp/3);
    DD=qq1+pp1;
end
qq2=sqrt(DD);
%Calculation of critical period of oscillation Tu
mm1=-qq/2+qq2; mm2=-qq/2-qq2;
nn1=mm1; nn2=mm2;
if nn1<0    nn1=abs(nn1);
end
if nn2<0    nn2=abs(nn2);
end
uu1=(nn1)^(1/3); vv1=(nn2)^(1/3);
if mm1<0    uu1=-uu1;
end
if mm2<0    vv1=-vv1;
end
zz1=uu1+vv1; beta=-((uu1+vv1)/2)-aa/3;
if beta>1    beta=.9999;
elseif beta<-1    beta=-1;
end
omegau=acos(beta)/T0; Tu=2*3.14/omegau;
if beta==0    Tu=4*T0;
elseif beta==-1 & zz1==-1    Tu=2*T0;
end
```

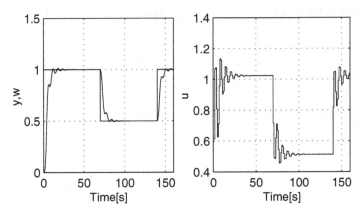

Figure 4.40. Example 4.5 – simulation results: control of model $G_D(z)$

$$u(k) = 1.6002[y(k-1) - y(k)] + 0.6135[w(k) - y(k)]$$
$$+1.4818[2y(k-1) - y(k-2) - y(k)] + u(k-1)$$

Simulation results for the model $G_D(z)$ are shown in Figure 4.40.

Example 4.6. Simulate control of the third-order model from Example 4.5 using the Takahashi controller (4.192), (4.193) with recursive identification (self-tuning controller). Choose the sampling period $T_0 = 1$ s, the initial parameter estimates as $\hat{\Theta}^T(0) = [0.1, 0.2, 0.3, 0.1, 0.2, 0.3]$ and the initial value of the directional forgetting factor set to $\varphi(0) = 1$.

Design a simulation program using the diagram in Figure 4.39 and Listing 4.3. Simulation results are shown in Figure 4.41.

It is clear from the recorded control responses in Figure 4.41 that in this case of a third-order self-tuning controller, the control quality is very good (without *a priori* information about controlled system parameters).

4.8 Simulation Examples in the Simulink® Environment

Simulation is a useful tool for the synthesis of control systems, allowing one not only to create mathematical models of a process but also to design virtual controllers in a computer. The mathematical models provided are sufficiently close to a real object that simulation can be used to verify the dynamic characteristics of control loops when the structure or parameters of the controller change. The models of the processes may also be excited by various random noise generators which can simulate the stochastic characteristics of process noise signals with similar properties to disturbance signals measured in the machinery. Simulation results are valuable for implementation of a chosen controller (control algorithm) under laboratory and industrial conditions. It must

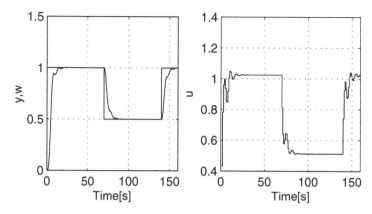

Figure 4.41. Example 4.6 – simulation results: self-tuning control of the third-order model $G_D(z)$

be borne in mind, however, that the practical application of a controller verified by simulation cannot be taken as a routine event. Obviously simulation and laboratory conditions can be quite different from those in real plants, and therefore one must verify its practicability with regard to process dynamics and the required standard of control quality (for example maximum allowable overshoot, accuracy, settling time, *etc.*).

While in Section 4.7 discrete models of the processes were used and the particular algorithms were programmed as M–Functions of the MATLAB® programming language, in this section simulations were performed using the Self-tuning Controllers Simulink Library [104] – see Section 7.1.

Only higher-order systems which have been approximated by second- or third-order models in the identification procedure have been chosen for simulation verification. Simulation verification has been limited to digital PID controllers based on the Ziegler–Nichols and the pole assignment methods, which have proved to be the best in practical applications. A scheme of the control circuit used for simulation in the MATLAB®/SIMULINK® environment is shown in Figure 4.42.

Note

The names of the particular controllers in the following examples and symbolic representations are given in Table 7.1, Section 7.1.1.

4.8.1 Simulation Control of Fourth-order Model

As an example of verification by computer simulation a fourth-order system with the transfer function

$$G(s) = \frac{1}{(s+1)^4} \tag{4.207}$$

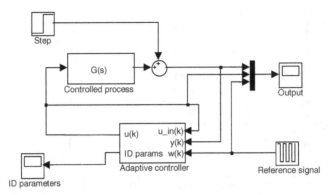

Figure 4.42. Control circuit used for simulation

was used. Simulation verification was realized using four types of controller included in the Self-tuning Controllers SIMULINK® Library [104] (see Section 7.1). In time interval $t = 150$–200 s a constant disturbance $v(k) = 0.5$ was injected onto the system output. The controller output value $u(k)$ was limited within the range $\langle 0; 2 \rangle$. The initial value of the directional forgetting factor was chosen as $\varphi(0) = 1$. The initial values of the model parameter estimates are $\hat{\Theta}^T(0) = [0.1, 0.2, 0.3, 0.4]$ for a second-order model, and $\hat{\Theta}^T(0) = [0.1, 0.2, 0.3, 0.4, 0.5, 0.6]$ for a third-order model.

Example 4.7. Consider a continuous-time controlled process with transfer function (4.207). Simulate the control loop using a PID Takahashi controller (*zn2tak*), choose sampling period $T_0 = 1$ s. Figure 4.43 illustrates the simulation control performance using Takahashi controller (4.38), (4.143). Critical parameters $K_{Pu}(T_0)$ and $T_u(T_0)$ were calculated recursively using the equations delivered in Section 4.7.4 for the identification of a second-order model.

Example 4.8. Consider a continuous-time controlled process with transfer function (4.207). Simulate the control loop using a PID Takahashi controller (*zn3tak*), choose sampling period $T_0 = 1.5$ s. Figure 4.44 illustrates the simulation control performance using Takahashi controller (4.38), (4.143). Critical parameters $K_{Pu}(T_0)$ and $T_u(T_0)$ were recursively calculated using the equations derived in Section 4.7.4 for the identification of a third-order model.

Example 4.9. Consider a continuous-time controlled process with transfer function (4.207). Simulate the control loop using a pole assignment controller PID–A1 (*pp2a_1*). Choose sampling period $T_0 = 2.5$ s, damping factor $\xi = 1$ and natural frequency $\omega_n = 0.3$. Figure 4.45 illustrates the simulation control performance using controller (4.120), (4.126), (4.127).

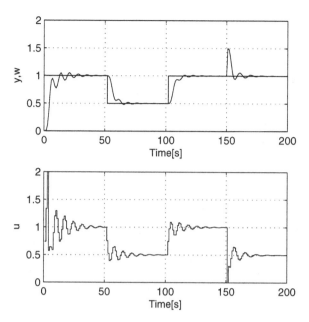

Figure 4.43. Example 4.7 – simulation control performance using a PID Takahashi controller (*zn2tak*)

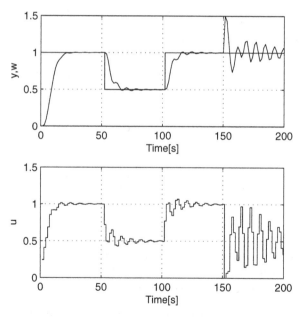

Figure 4.44. Example 4.8 – simulation control performance using a PID Takahashi controller (*zn3tak*)

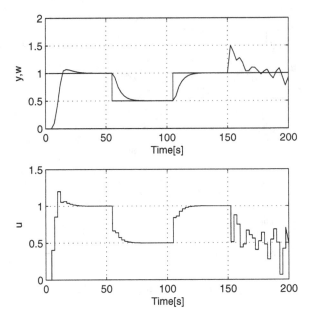

Figure 4.45. Example 4.9 – simulation control performance using a pole assignment controller PID–A1 (*pp2a_1*)

Example 4.10. Consider a continuous-time controlled process with transfer function (4.207). Simulate the control loop using a pole assignment controller PID–B1 (*pp2b_1*). Choose sampling period $T_0 = 2.5$ s, damping factor $\xi = 1$ and natural frequency $\omega_n = 0.11$. Figure 4.46 illustrates simulation control performance using controller (4.126), (4.133), (4.138), (4.139).

4.8.2 Simulation Control of Third-order Nonminimum Phase Model

As an example of verification by computer simulation a fourth-order system with the transfer function

$$G(s) = \frac{1 - 2s}{(s+1)^3} \tag{4.208}$$

was used. Simulation verification was realized using the same four types of controller as in Section 4.8.1. Simulation conditions ($v(k), \hat{\Theta}^T(0)$ and limiting of $u(k)$) were also chosen to be the same.

Example 4.11. Consider the continuous-time transfer function (4.208) as a model of a controlled process. Simulate the control loop using a PID Takahashi controller (*zn2tak*), choose sampling period $T_0 = 1$ s. Figure 4.47 illustrates the simulation control performance using Takahashi controller (4.38), (4.143).

Critical parameters $K_{Pu}(T_0)$ and $T_u(T_0)$ were calculated recursively using the equations derived in Section 4.7.4 for the identification of a second-order model.

Example 4.12. Consider a continuous-time controlled process with transfer function (4.208). Simulate the control loop using a PID Takahashi controller (*zn3tak*), choose sampling period $T_0 = 0.35$ s. Figure 4.48 illustrates the simulation control performance using Takahashi controller (4.38), (4.143). Critical parameters $K_{Pu}(T_0)$ and $T_u(T_0)$ were calculated recursively using the equations derived in Section 4.7.4 for the identification of a third-order model.

Example 4.13. Consider a continuous-time controlled process with transfer function (4.208). Simulate the control loop using a pole assignment controller PID–A1 (*pp2a_1*). Choose sampling period $T_0 = 2.5$ s, damping factor $\xi = 1$ and natural frequency $\omega_n = 0.11$. Figure 4.49 illustrates the simulation control performance using controller (4.120), (4.126), (4.127).

Example 4.14. Consider a continuous-time controlled process with transfer function (4.208). Simulate the control loop using a pole assignment controller PID–B1 (*pp2b_1*). Choose sampling period $T_0 = 2.5$ s, damping factor $\xi = 1$ and natural frequency $\omega_n = 0.11$. Figure 4.50 illustrates the simulation control performance using controller (4.126), (4.133), (4.138), (4.139).

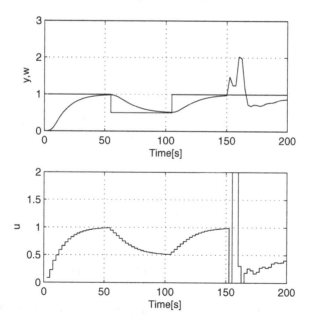

Figure 4.46. Example 4.10 – simulation control performance using a pole assignment controller PID–B1 (*pp2b_1*)

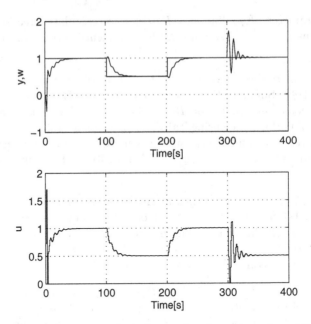

Figure 4.47. Example 4.11 – simulation control performance using a PID Takahashi controller (*zn2tak*)

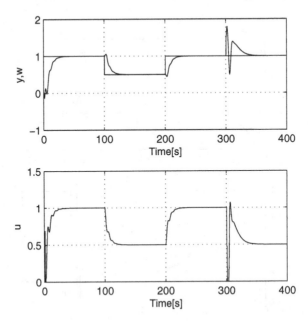

Figure 4.48. Example 4.12 – simulation control performance using a PID Takahashi controller (*zn3tak*)

The purpose of the examples above was not to find optimal values of the particular variables in the control loops; this task must be solved by a user according to his control demands. The analysis and discussion of simulation results with regard to the chosen initial conditions and input parameters can be left to the reader as a problem to solve.

4.9 Summary of chapter

PID controllers are still the most widely used controllers in the industry and thus corresponding attention is paid to them in this chapter. The purpose of this part is not only to give an overview of digital PID controllers suitable for self-tuning modifications but attention is also focused on the implementation of controllers in industrial application. The chapter is supplemented by examples of the design of digital PID controllers and listings of their programs. Self-tuning versions of these controllers can then be programmed by incorporating recursive identification, and used for both simulation purposes and implementations in real time. The dynamic behaviour of these types of controllers is determined by a number of simulation verifications.

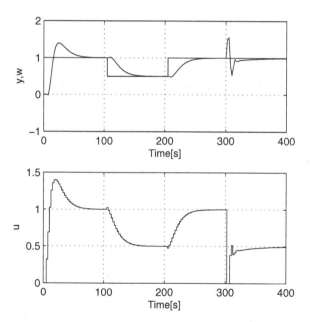

Figure 4.49. Example 4.13 – simulation control performance using a pole assignment controller PID–A1 (*pp2a_1*)

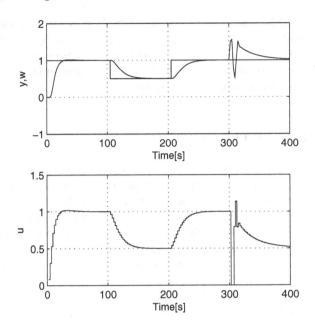

Figure 4.50. Example 4.14 – simulation control performance using a pole assignment controller PID–B1 (*pp2b_1*)

Problems

4.1. Design PID–B1 type controllers suitable for the control of processes described by the models used in Example 4.1 Choose a suitable pole assignment, simulate the control, and compare the simulation results achieved with the results in the above mentioned Example 4.1.

4.2. Design PID–A2 type controllers for control of processes described by the models used in Example 4.1. Simulate the control and compare the simulation results achieved with the results in Example 4.1.

4.3. Modify the PID–B1 type controllers designed in Problem 4.1 as self-tuning controllers.

4.4. Verify by experiment the influence of the damping factor ξ and natural frequency ω_n on control by self-tuning PID type controllers based on the pole assignment method.

4.5. Verify by experiment the influence of the sampling period T_0 on control by self-tuning PID type controllers based on the pole assignment method.

4.6. Verify by experiment the influence of the sampling period T_0 on control by self-tuning PID type controllers based on the modified Ziegler–Nichols method.

4.7. Verify by experiment the influence of random disturbances on the behaviour of closed loop systems with self-tuning PID type controllers based on the modified Ziegler–Nichols method.

5

Algebraic Methods for Self-tuning Controller Design

Different forms of self-tuning controllers may be suggested according to the type of plant model chosen (and consequently, the chosen identification method), according to quality criteria, or according to the mathematical procedure used during the derivation of controller equations. Different algebraic control theory-based algorithms are introduced in this chapter. These algorithms use various criteria for the process control – *i.e.* the dead-beat method, the pole assignment method, and the linear quadratic (LQ) control method.

This chapter is further divided into five sections: Section 5.1 summarizes the basic starting assumptions for controller tuning using algebraic methods; Section 5.2 considers a description of the dead-beat method – in both its strong and weak versions; Section 5.3 is concerned with the pole assignment method; Section 5.4 solves the problems associated with LQ criterion minimization; and Section 5.5 contains illustrative examples and concludes this chapter.

5.1 Basic Terms

Let us consider validation of the following assumptions:

- The controller equations are expressed in discrete form.
- The plant has, for a zero disturbance signal v, the ARMAX model

$$A(z^{-1})y(k) = z^{-d}B(z^{-1})u(k) + C(z^{-1})e_s(k) \qquad (5.1)$$

where

$$A(z^{-1}) = 1 + a_1 z^{-1} + \ldots + a_{na} z^{-na}$$
$$B(z^{-1}) = b_1 z^{-1} + \ldots + b_{nb} z^{-nb}$$
$$C(z^{-1}) = 1 + c_1 z^{-1} + \ldots + c_{nc} z^{-nc}$$

d is a time delay expressed as an integer multiple of sampling period T_0, $e_s(k)$ is a sequence of non-correlative noise with zero mean value.

This plant model is the result of recursive identification in the self-tuning controller algorithm. It is possible to obtain simpler forms of the model (5.1) for $C(z^{-1}) = 1$ (ARX model), for $d = 0$ (zero time delay) and for $na = nb = n$ (equal degree of polynomials).

- Process input and output are in an incremental form.
- The plant has zero initial conditions.

The controller design will be based on the general block scheme of a closed loop controller with two degrees of freedom (see Figure 5.1). The influence of changes in the reference value w and the influence of disturbances v placed on the input of a plant can be observed.

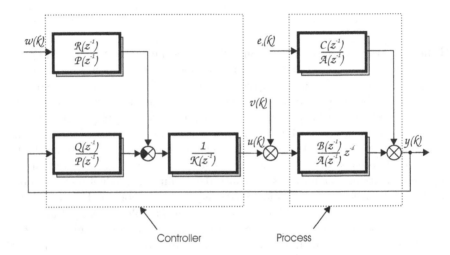

Figure 5.1. Block scheme of a control process with a controller with two degrees of freedom

Given that $v(k) = 0$, and $e_s(k) = 0$, the following controller equation can be extrapolated:

$$P(z^{-1})K(z^{-1})u(k) = R(z^{-1})w(k) - Q(z^{-1})y(k) \qquad (5.2)$$

This equation could be simplified for $K(z^{-1}) = 1$. A special case is a controller with only one degree of freedom (Figure 5.2), which is valid for $R(z^{-1}) = Q(z^{-1})$, and which works with a tracking error $e(k) = w(k) - y(k)$. It can be proved that such a controller is only sub-optimal for tasks with reference value tracking.

A similar procedure could be used to determine a controller for a nonzero disturbance $v(k)$ and with zero reference value $w(k)$:

$$P(z^{-1})K(z^{-1})u(k) = P(z^{-1})K(z^{-1})v(k) - Q(z^{-1})y(k) \qquad (5.3)$$

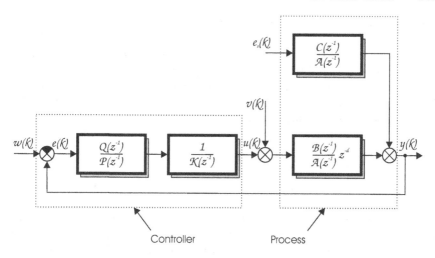

Figure 5.2. Block scheme of a control process with a controller with one degree of freedom

From Equations (5.1) and (5.2), it is possible to determine the transfer function of a closed loop, thus:

$$G_w(z) = \frac{Y(z)}{W(z)} = \frac{B(z^{-1})R(z^{-1})}{A(z^{-1})K(z^{-1})P(z^{-1}) + B(z^{-1})Q(z^{-1})} \qquad (5.4)$$

and from Equations (5.1) and (5.3) the transfer function from v to y, thus:

$$G_v(z) = \frac{Y(z)}{V(z)} = \frac{B(z^{-1})K(z^{-1})P(z^{-1})}{A(z^{-1})K(z^{-1})P(z^{-1}) + B(z^{-1})Q(z^{-1})} \qquad (5.5)$$

The most important influence on a control process (mainly on its stability) is the denominator of the transfer function. The transfer function of a closed loop is suitably adjusted by choosing a controller transfer function to fit the desired criterion.

Above all, the controller must be designed to achieve closed loop stability. One can define the stability condition as the requirement for the existence of only stable pole, which means

$$|z_i| < 1, \qquad i = 1, 2, \ldots n \qquad (5.6)$$

The controlled system (plant) itself could be unstable (it has unstable polynomial $A(z^{-1})$), or nonminimum phase (unstable polynomial $B(z^{-1})$). Some controllers can successfully control these systems, while some produce instability, and thus cannot be used. The problem is that unstable plant polynomials may be cancelled by controller polynomials. Formally, it is possible to eliminate these polynomials, but the instability of the closed loop remains the same. The instability can lead to stabilization of the controlled variable, while the controller output oscillates. The problem of a polynomial cancelling unstable polynomials is that the plant model can never describe the behaviour of a plant perfectly – and the controller, designed according to the plant model, cannot fully compensate the sources of instability in a real plant. For instance, in order to cancel first-degree unstable polynomials ($|a_1| > 1$), which is different for arbitrary small Δa_1, the result is not close to 1, but rather, the infinite divergent sequence

$$(1 + a_1 z^{-1}) : \left[1 + (a_1 + \Delta a_1)z^{-1}\right] =$$
$$1 - \Delta a_1 z^{-1} + a_1 \Delta a_1 z^{-2} - a_1^2 \Delta a_1 z^{-3} + a_1^3 \Delta a_1 z^{-4} + \ldots$$

after omitting the products $\Delta a_1 \Delta a_1$. To avoid these cancelling problems, factorization of polynomials is sometimes performed for some methods, which divides the polynomials into a stable part (denoted with the index +) and an unstable part (-), for example

$$A(z^{-1}) = A^+(z^{-1})A^-(z^{-1}) \tag{5.7}$$

which cancels only the stable parts of polynomials. Another way is by spectral factorization of polynomials, which leaves the stable parts unchanged and changes unstable parts into reciprocal ones, and by this operation making them stable.

Algebraic Control Theory is based on the methods of linear algebra [105, 106, 107]. The basic tool for the description of transfer functions is the application of polynomials, expressed as a finite sequence of numbers – *i.e.* the coefficients of a polynomial. Thus, signals are expressed as an infinite sequence of numbers. Controller synthesis consists in solving linear polynomial (Diophantine) equations of the general form [108]

$$AX + BY = C \tag{5.8}$$

The equation can be solved if the common divisor of polynomials A and B also divides the polynomial C. The Equation (5.8) has an infinite number of solutions. If a particular solution is X_0, Y_0, then the general solution has the form

$$X = X_0 + BT$$
$$Y = Y_0 - AT \tag{5.9}$$

where T is an arbitrary polynomial. Hence, the choice of polynomial T could fulfil additional requirements on the solution of the equation. Diophantine equations can be solved using the uncertain coefficients method – which is based on comparing coefficients of the same power. The Diophantine equation is then transformed into a system of linear algebraic equations. It is also possible to solve Diophantine equations using the function *axbyc* from the MATLAB® Polynomial Toolbox [109] during simulations.

5.2 Dead-beat Method

There exist both strong and a weak versions of this method – if a process output and a reference output are the same for an arbitrary time point, or if this equality is valid only for the time points of the sampling period. This causes a difference in a controller output progression, which does achieve (after a change) the steady state (in the strong version), or where it only converges towards the steady state (weak version).

5.2.1 Strong Version of the Dead-beat Method

An algorithm can be derived [110, 111] for a control loop according to Figure 5.1 with the following simplifications: there is no noise signal nor disturbance, the system does not consist of a time-delay and the polynomial $K(z^{-1}) = 1$. The Equations (5.1) and (5.2), which describe a controlled system and a controller, have, after their transformation, the form

$$Y(z^{-1}) = \frac{B(z^{-1})}{A(z^{-1})} U(z^{-1}) \tag{5.10}$$

$$U(z^{-1}) = \frac{R(z^{-1})}{P(z^{-1})} W(z^{-1}) - \frac{Q(z^{-1})}{P(z^{-1})} Y(z^{-1}) \tag{5.11}$$

If $U(z^{-1})$ from (5.11) is substituted into (5.10), then after adjustment

$$Y(z^{-1}) = \frac{B(z^{-1})R(z^{-1})}{A(z^{-1})P(z^{-1}) + B(z^{-1})Q(z^{-1})} W(z^{-1}) \tag{5.12}$$

Using the opposite operation, if $Y(z^{-1})$ is omitted from (5.11), one can obtain the expression for an input signal

$$U(z^{-1}) = \frac{A(z^{-1})R(z^{-1})}{A(z^{-1})P(z^{-1}) + B(z^{-1})Q(z^{-1})} W(z^{-1}) \tag{5.13}$$

The polynomial for the tracking error is, after substituting from (5.12) for the

$$\begin{aligned} E(z^{-1}) &= W(z^{-1}) - Y(z^{-1}) \\ &= \left[1 - \frac{B(z^{-1})R(z^{-1})}{A(z^{-1})P(z^{-1}) + B(z^{-1})Q(z^{-1})} \right] W(z^{-1}) \end{aligned} \tag{5.14}$$

If the requirement is the achievement of zero level tracking error after a change to the controller output in a finite number of control steps (dead-beat), the polynomial $E(z^{-1})$ must be as simple as possible. This condition is valid only if this polynomial is not in the form of a fraction, which means if

$$A(z^{-1})P(z^{-1}) + B(z^{-1})Q(z^{-1}) = 1 \tag{5.15}$$

is valid. Equation (5.15) has a minimal solution if the following equations

$$\partial P(z^{-1}) = \partial B(z^{-1}) - 1$$
$$\partial Q(z^{-1}) = \partial A(z^{-1}) - 1$$

(5.16)

are valid for the degrees of polynomials $P(z^{-1})$ and $Q(z^{-1})$, and if the polynomials $A(z^{-1})$ and $B(z^{-1})$ are coprime. The expression (5.15) is consequently the condition of closed loop stability. Equation (5.14) is, after application of the condition (5.15), reduced to

$$E(z^{-1}) = \left[1 - B(z^{-1})R(z^{-1})\right] W(z^{-1})$$

(5.17)

The sequence $W(z^{-1})$, which describes the time progression of reference value $w(k)$, could be expressed as the ratio of polynomials, as follows:

$$W(z^{-1}) = \frac{N_w(z^{-1})}{D_w(z^{-1})}$$

(5.18)

Consequent simplification of the polynomial $E(z^{-1})$ is possible if the polynomial $D_w(z^{-1})$ divides the expression $1 - B(z^{-1})R(z^{-1})$. This ratio will be denoted as polynomial $S(z^{-1})$, thus

$$S(z^{-1}) = \frac{1 - B(z^{-1})R(z^{-1})}{D_w(z^{-1})}$$

(5.19)

and Equation (5.19) is adjusted to the form

$$D_w(z^{-1})S(z^{-1}) + B(z^{-1})R(z^{-1}) = 1$$

(5.20)

Similar to the solution of Equation (5.15), this equation has a minimal solution if the following condition for degrees of polynomials is valid:

$$\partial R(z^{-1}) = \partial D_w(z^{-1}) - 1$$
$$\partial S(z^{-1}) = \partial B(z^{-1}) - 1$$

(5.21)

Polynomial $S(z^{-1})$ serves as a by-product and there is no need to compute it in the course of tuning the controller. It could be used for computing a tracking error

$$E(z^{-1}) = S(z^{-1})N_w(z^{-1})$$

(5.22)

as in the results from Equations (5.17) to (5.19).

Polynomial Equations (5.15) and (5.20) are usually solved using the uncertain coefficient method (the comparison of elements with the same power of z). The above-mentioned algorithm is used for reference signal tracking, whose progression must be known in order to tune a controller. Step changes of the reference signal are usually used in practice. The step function with size w_1 can be expressed as

$$W(z^{-1}) = \frac{N_w(z^{-1})}{D_w(z^{-1})} = \frac{w_1}{1 - z^{-1}} \quad (5.23)$$

Because it is simpler, let us consider unitary change of the reference ($w_1 = 1$); Equation (5.20) is then simplified to a form where

$$(1 - z^{-1})S(z^{-1}) + B(z^{-1})R(z^{-1}) = 1 \quad (5.24)$$

With respect to Equation (5.21), the polynomial $R(z^{-1})$ has zero degree and the solution of Equation (5.24) is the relationship

$$R(z^{-1}) = r_0 = \frac{1}{b_1 + b_2 + \ldots + b_n} \quad (5.25)$$

Hence, the controller design is simplified to computation of the coefficients of the polynomials $P(z^{-1})$ and $Q(z^{-1})$ from Equation (5.15) and the coefficient r_0 from Equation (5.25). The tracking error polynomial for step changes of a reference signal is

$$E(z^{-1}) = S(z^{-1}) = \left[1 - B(z^{-1})r_0\right]\frac{1}{1 - z^{-1}} \quad (5.26)$$

and it has n nonzero coefficients. The sequence of values of the process output is described by the equation

$$Y(z^{-1}) = W(z^{-1}) - E(z^{-1}) = B(z^{-1})r_0\frac{1}{1 - z^{-1}} \quad (5.27)$$

which could be, after dividing the polynomials, written as

$$Y(z^{-1}) = r_0\left[b_1 z^{-1} + (b_1 + b_2)z^{-2} + \ldots + (b_1 + b_2 + \ldots + b_n)z^{-n}\right. \\ \left. + (b_1 + b_2 + \ldots + b_n)z^{-(n+1)} + \ldots\right] \quad (5.28)$$

Beginning with the n-th step, the coefficients of the sequence are equal to 1, which means: $y(k) = w(k)$. The controller output $U(z^{-1})$ is, according to Equations (5.10) and (5.27) equal to

$$U(z^{-1}) = \frac{A(z^{-1})}{B(z^{-1})}Y(z^{-1}) = A(z^{-1})r_0\frac{1}{1 - z^{-1}} \quad (5.29)$$

and after extension

$$U(z^{-1}) = r_0\left[1 + (1 + a_1)z^{-1} + \ldots + (1 + a_1 + a_2 + \ldots + a_n)z^{-n}\right. \\ \left. + (1 + a_1 + a_2 + \ldots + a_n)z^{-(n+1)} + \ldots\right] \quad (5.30)$$

Controller output values are constant from the n-th step, and they are equal to the reciprocal value of system gain. The transfer function of a closed loop for step changes of a reference signal can be derived from Equations (5.23) and (5.27)

$$G_w(z) = \frac{Y(z)}{W(z)} = B(z^{-1})r_0 \tag{5.31}$$

After transformation of the transfer function to a variable of z, the denominator is equal to z^n and the characteristic equation has n multiple zero roots. Zero poles ensure the fastest stabilization of a control process.

If a system contains a time delay d, the degree of the polynomial $B(z^{-1})$ is increased to $n + d$

$$B(z^{-1}) = b_1 z^{-(1+d)} + b_2 z^{-(2+d)} + \ldots + b_n z^{-(n+d)}$$

The procedure for controller design remains the same – tracking of the reference signal is performed for $n + d$ sampling periods. The characteristic equation of a closed loop transfer function has a pole with $(n + d)$-th power.

Note
It is possible to obtain the simplest tracking error polynomial for the known progression of a reference signal if the condition equations are in the form

$$A(z^{-1})P(z^{-1}) + B(z^{-1})Q(z^{-1}) = N_w(z^{-1})$$

$$D_w(z^{-1})S(z^{-1}) + B(z^{-1})R(z^{-1}) = N_w(z^{-1})$$

Such a controller meets the requirement for a finite number of steps only for zero initial conditions, but the controller from Equations (5.15) and (5.20) is more general and works with arbitrary initial conditions on the process output, controller output and reference signal.

The algorithm for the dead-beat method was designed for reference tracking, and it does not consist of an integral (or summing in a discrete form) action. Thus, it is not convenient for disturbance elimination or for a controller with one degree of freedom because it causes steady-state errors. Integral behaviour of a controller can be achieved if it consists of the element $K(z^{-1}) = 1 - z^{-1}$. From the transfer function of closed loop (5.4), the modified conditional Equation (5.15) can be derived in the form

$$A(z^{-1})K(z^{-1})P(z^{-1}) + B(z^{-1})Q(z^{-1}) = 1 \tag{5.32}$$

where the degree of the polynomial $Q(z^{-1})$ is increased by 1

$$\partial Q(z^{-1}) = \partial A(z^{-1}) + \partial K(z^{-1}) - 1 = \partial A(z^{-1}). \tag{5.33}$$

The condition (5.20) for computing the polynomial $R(z^{-1})$ does not change and the element $K(z^{-1})$ does not have any influence on tracking the reference signal.

The design of controller parameters for a controller with two degrees of freedom and with the element $K(z^{-1})$ will be demonstrated in the example on control of a second-order system with a transfer function

$$G_P(z) = \frac{B(z^{-1})}{A(z^{-1})} = \frac{b_1 z^{-1} + b_2 z^{-2}}{1 + a_1 z^{-1} + a_2 z^{-2}}$$

The degree of the polynomial $P(z^{-1})$ is, according to (5.16) $\partial P(z^{-1}) = 1$ and the degree of the polynomial $Q(z^{-1})$ with respect to (5.33), $\partial Q(z^{-1}) = 2$. Equation (5.32) is then expressed in the form

$$(1 + a_1 z^{-1} + a_2 z^{-2})(1 - z^{-1})(p_0 + p_1 z^{-1})$$

$$+(b_1 z^{-1} + b_2 z^{-2})(q_0 + q_1 z^{-1} + q_2 z^{-2}) = 1$$

The uncertain coefficients method leads to a system of nonlinear equations $(p_0 = 1)$

$$\begin{bmatrix} b_1 & 0 & 0 & 1 \\ b_2 & b_1 & 0 & a_1 - 1 \\ 0 & b_2 & b_1 & a_2 - a_1 \\ 0 & 0 & b_2 & -a_2 \end{bmatrix} \begin{bmatrix} q_0 \\ q_1 \\ q_2 \\ p_1 \end{bmatrix} = \begin{bmatrix} 1 - a_1 \\ a_1 - a_2 \\ a_2 \\ 0 \end{bmatrix}$$

and their solution is as follows

$$q_2 = \frac{b_1 a_2^2 - b_2 a_2 \left[a_1 - a_2 + \frac{b_2}{b_1}(a_1 - 1) \right]}{(a_2 - a_1)b_1 b_2 + b_2^2(1 - a_2 + \frac{b_2}{b_1}) + a_2 b_1^2}$$

$$q_1 = \frac{a_1 - a_2}{b_1} + \frac{b_2}{b_1^2}(a_1 - 1) + \frac{b_2 q_2}{a_2 b_1}(1 - a_1 + \frac{b_2}{b_1})$$

$$q_0 = \frac{a_2(1 - a_1) - b_2 q_2}{a_2 b_1}$$

$$p_1 = \frac{b_2 q_2}{a_2}$$

Polynomial $R(z^{-1})$ has, according to Equation (5.25), only one parameter if only step changes of the reference signal are performed

$$r_0 = \frac{1}{b_1 + b_2}$$

Computed parameters are then substituted into the controller equation, as derived from Equation (5.3)

$$u(k) = r_0 w(k) - q_0 y(k) - q_1 y(k-1) - q_2 y(k-2) + (1 - p_1)u(k-1) + p_1 u(k-2)$$

Examples of control processes with this type of controller (denoted **DB1**) are given in Section 5.5.

The controller with one degree of freedom, where $R(z^{-1}) = Q(z^{-1})$, contains an input tracking error $e(k)$, but the polynomial parameters do not change

$$u(k) = q_0 e(k) + q_1 e(k-1) + q_2 e(k-2) + (1 - p_1)u(k-1) + p_1 u(k-2)$$

This type of controller does not ensure tracking of the reference signal with zero error. It results from Equation (5.17) for the tracking error sequence, which need not be finite.

The common disadvantage of the dead-beat method is that for fast stabilization of the process output, it is necessary to produce large peaks at the controller output – especially in the first step. During decreases in the sampling period, the controller output must increase to ensure stabilization of the process output in a shorter time. Practically realized actuators have a fixed range and in the first steps, the process input may be saturated, which causes degradation in the quality of control (an increase in the number of steps to stabilization of the process output). The simplest way to decrease controller output peaks is to increase the sampling period, or to filter the step changes of the reference.

5.2.2 Weak Version of the Dead-beat Method

A controller for the weak version of the dead-beat method may be derived from the condition that the process output tracks changes of the reference signal with a delay of one sampling period $y(k) = w(k-1)$. Thus, the transfer function of the closed loop is equal to $G_w(z) = z^{-1}$. Using Equation (5.10), it can be computed as follows

$$y(k) = B(z^{-1})u(k) - \left[A(z^{-1}) - 1\right] y(k) \tag{5.34}$$

and, after simplifying, $y(k) = w(k-1)$, and after a shift of a single sampling period, it is possible to obtain the following equation for the controller

$$u(k) = \frac{1}{b_1} \left[\sum_{i=1}^{n} a_i y(k-i+1) + w(k) - \sum_{i=2}^{n} b_i u(k-i+1) \right] \tag{5.35}$$

It is evident from the form of this equation that the controller has two degrees of freedom, and the polynomial $K(z^{-1}) = 1$. For a second-order system, the controller (5.35) has the form

$$u(k) = \frac{1}{b_1} w(k) + \frac{a_1}{b_1} y(k) + \frac{a_2}{b_1} y(k-1) - \frac{b_2}{b_1} u(k-1)$$

and this will be denoted **DB2**.

It is possible, by comparing the general form of controller according to (5.2), to write Equation (5.35) using polynomials, as follows

$$P(z^{-1}) = zB(z^{-1}) = b_1 + b_2 z^{-1} + \ldots + b_n z^{-(n-1)}$$

$$Q(z^{-1}) = z \left[1 - A(z^{-1}) \right] = -a_1 - a_2 z^{-1} - \ldots - a_n z^{-(n-1)}$$

$$R(z^{-1}) = 1$$

$$K(z^{-1}) = 1$$

And after substituting the polynomials into the transfer function of the closed loop (5.4), as

$$G_w(z) = \frac{Y(z)}{W(z)} = \frac{B(z^{-1})}{A(z^{-1})zB(z^{-1}) + B(z^{-1})z\left[1 - A(z^{-1})\right]} = \frac{B(z^{-1})}{zB(z^{-1})} \tag{5.36}$$

This transfer function can be simplified to the original form z^{-1} only under the assumption that the polynomial $B(z^{-1})$ is stable. Thus, the method is not suitable for the control of nonminimum phase systems, because the cancelling of an unstable polynomial $B(z^{-1})$ leads to an unstable control process.

Equal polynomials could be derived from the conditional equation

$$A(z^{-1})P(z^{-1}) + B(z^{-1})Q(z^{-1}) = zB(z^{-1}) \tag{5.37}$$

The stability of nonminimum phase systems is ensured by factorization of polynomial $B(z^{-1})$ according to Equation (5.7). The only stable part of this polynomial is located on the right side of Equation (5.37)

$$A(z^{-1})P(z^{-1}) + B(z^{-1})Q(z^{-1}) = zB^+(z^{-1}) \tag{5.38}$$

and only these stable parts are cancelled. This results in the closed loop transfer function $G_w(z) = z^{-1}B^-(z^{-1})$.

As was mentioned at the beginning of this chapter, there is equality between the reference and the process output only during the sampling period. The course of the controller output is given by the ratio of the coprime polynomials, so the controller output does not achieve steady state. The process output oscillates with a period equal to two sampling periods around the reference value, as will be demonstrated in the example given in Section 5.5.

5.3 Pole Assignment Method

A controller based on the pole assignment (placement) of a closed control loop is designed to achieving the pre-set poles of the characteristic polynomial [69]. Besides the requirement for stability, it is possible, using suitable pole assignment, to obtain the required characteristic of the output variable of the closed loop – for instance maximal overshoot, damping, *etc.* The chosen types of controllers based on pole assignment methods have already been cited in Section 4.7.3. In this section, they will be generalized and completed.

5.3.1 Effects of Pole Assignment on the Control Process

The relationship between the position of the poles and the course of a control process will be demonstrated with the 2-nd order transfer function of the form

$$G_w(s) = \frac{\omega_n^2}{s^2 + 2\xi\omega_n s + \omega_n^2} \tag{5.39}$$

where ξ is the damping factor, and ω_n is the natural frequency of oscillation. The real damping of the system is equal to $\xi\omega_n$ (damping factor), and the real (damped) oscillating frequency of the system is $\omega_n\sqrt{1-\xi^2}$. The characteristic equation of the transfer function (5.39) is

$$s^2 + 2\xi\omega_n s + \omega_n^2 = 0$$

which has poles (roots)

$$s_{1,2} = -\xi\omega_n \pm \omega_n\sqrt{\xi^2 - 1} \tag{5.40}$$

To achieve stability of the control process, the poles have to lie in the left half-plane of the plane s, and the conditions $\xi > 0$ and $\omega_n > 0$ must be valid. An oscillating damped control process will be obtained, if the roots are complex-conjugate – i.e., if the damping factor is in the range $0 < \xi < 1$. In this case, the response of the system to a unit step has the form [112]

$$y(t) = 1 - \frac{e^{-\xi\omega_n t}}{\sqrt{1-\xi^2}} \left[\xi\sin\left(\omega_n\sqrt{1-\xi^2}t\right) + \sqrt{1-\xi^2}\cos\left(\omega_n\sqrt{1-\xi^2}t\right) \right] \tag{5.41}$$

From Equation (5.41), the oscillation frequency is $\omega_n\sqrt{1-\xi^2}$ and accordingly the oscillation period is given by

$$T_k = \frac{2\pi}{\omega_n\sqrt{1-\xi^2}} \tag{5.42}$$

The first (maximal) overshoot of the process output will be observed in a time equal to half the period T_k. The size of the process output at this time point is computed by setting $t = T_k/2$ in Equation (5.41)

$$y_{max} = y\left(\frac{T_k}{2}\right) = 1 + e^{-\frac{\xi\pi}{\sqrt{1-\xi^2}}} \tag{5.43}$$

Thus, the maximum percentage overshoot of Δy is given by

$$\Delta y = 100 e^{-\frac{\xi\pi}{\sqrt{1-\xi^2}}}$$

The real roots ($\xi \geq 1$) represent reciprocal values of the time constants of the transfer function and lead to an aperiodical control process.

It is possible using the above analysis to choose the parameters ξ and ω_n so that the course of the control process fits the required properties.

Example 5.1. Compute the oscillation period and overshoot of the process output of transfer function (5.39) (the damping factor $\xi = 0.4$ and the natural frequency $w_n = 0.3$ s^{-1}) and compare with the course of the step response of this system given on Figure 5.3.

From Equation (5.42), the oscillation period is $T_k = 22.9$ s. The first overshoot will be observed in a time 11.45 s and its size, according to Equation (5.43) is $y_{max} = 1.25$. The computed values correspond to the values in Figure 5.3.

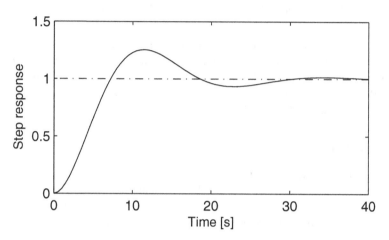

Figure 5.3. Step response of an underdamped second-order process (solid line – process output y, dash/dot line – reference value w)

After conversion of the transfer function (5.39) into a discrete form, the denominator of the transfer function contains the second-degree polynomial $D(z^{-1})$ in the form

$$D(z^{-1}) = 1 + d_1 z^{-1} + d_2 z^{-2}. \tag{5.44}$$

The roots of the continuous-time transfer function are converted into discrete form according to the relationship $z_i = \exp(s_i T_0)$, $i = 1, 2$; and the coefficients of the polynomial $D(z^{-1})$ then have the form

$$d_1 = -2 \exp(-\xi w_n T_0) \cos\left(w_n T_0 \sqrt{1 - \xi^2}\right); \quad \text{for } \xi \leq 1$$
$$d_1 = -2 \exp(-\xi w_n T_0) \cosh\left(w_n T_0 \sqrt{\xi^2 - 1}\right); \quad \text{for } \xi > 1 \tag{5.45}$$
$$d_2 = \exp(-2\xi w_n T_0)$$

The pole assignment of the discrete transfer function depends on the chosen sampling period T_0.

Below are some options for pole assignment of the closed loop discrete transfer function:

- for a control process with overshoot, the pair of complex-conjugate poles with positive real parts is chosen;
- for an aperiodical control process, the real poles are placed on the positive axis;
- for a dead-beat control process, the poles are placed at the zeros (see Section 5.2);
- the poles are derived from the transfer function of the controlled system; for example by using spectral factorization of the numerator and denominator (see Section 5.4).

The above-mentioned approaches can be mutually combined.

It is necessary to realize that it is not possible "to force" the controlled system to track the required behaviour with pole assignment. For example, a controller cannot eliminate time delay in the system, or the undershoot of a nonminimum phase system.

5.3.2 Algorithm Derivation

The derivation of equations for computing controller parameters is similar to the procedure with the dead-beat method. In the conditional Equation (5.15), the one on the right side of the equation – which places the poles to zero, is replaced by the polynomial $D(z^{-1})$ with chosen poles, thus

$$A(z^{-1})P(z^{-1}) + B(z^{-1})Q(z^{-1}) = D(z^{-1}) \qquad (5.46)$$

Likewise, Equation (5.32) is changed using the polynomial $K(z^{-1})$, as follows

$$A(z^{-1})K(z^{-1})P(z^{-1}) + B(z^{-1})Q(z^{-1}) = D(z^{-1}) \qquad (5.47)$$

If the following equation is valid

$$\partial D(z^{-1}) \leq \partial A(z^{-1}) + \partial B(z^{-1}) + \partial K(z^{-1}) - 1 \qquad (5.48)$$

then the relationships for determining the minimal degree of polynomials $P(z^{-1})$ and $Q(z^{-1})$ are equal those ones used according to the criterion of the dead-beat method – see Equations (5.16) and (5.33). If the condition (5.48) is not valid, the degrees of polynomials $P(z^{-1})$ and $Q(z^{-1})$ cannot be uniquely determined. For instance, for Equation (5.46), the number of determined parameters is equal to $\partial P(z^{-1}) + \partial Q(z^{-1}) + 2$, because a polynomial of n-th degree has $n + 1$ parameters. The number of equations, which are obtained by comparing elements of the same power, is either $\partial A(z^{-1}) + \partial P(z^{-1}) + 1$ or $\partial B(z^{-1}) + \partial Q(z^{-1}) + 1$ – where the greater value has to be chosen. Either the equation

$$\partial P(z^{-1}) + \partial Q(z^{-1}) + 2 = \partial A(z^{-1}) + \partial P(z^{-1}) + 1 \qquad (5.49)$$

or the equation

$$\partial P(z^{-1}) + \partial Q(z^{-1}) + 2 = \partial B(z^{-1}) + \partial Q(z^{-1}) + 1 \qquad (5.50)$$

is valid.

The solution of the first variant: from Equation (5.49) we have

$$\partial Q(z^{-1}) = \partial A(z^{-1}) - 1 \qquad (5.51)$$

and, on the condition that the first element on the left side of Equation (5.46) has a higher degree than the second one, the degree of polynomial $P(z^{-1})$ is determined as

$$\partial P(z^{-1}) = \partial D(z^{-1}) - \partial A(z^{-1}) \qquad (5.52)$$

In the opposite case, *i.e.* from Equation (5.50), the polynomial $P(z^{-1})$ has the degree

$$\partial P(z^{-1}) = \partial B(z^{-1}) - 1 \qquad (5.53)$$

and from Equation (5.46), the degree of polynomial $Q(z^{-1})$ is

$$\partial Q(z^{-1}) = \partial D(z^{-1}) - \partial B(z^{-1}) \qquad (5.54)$$

For example, where the polynomials $A(z^{-1})$ and $B(z^{-1})$ for the second-order system are of 2-nd degree and the polynomial $D(z^{-1})$ are of 3-rd degree, the condition (5.48) is fulfilled. The degrees of polynomials $P(z^{-1})$ and $Q(z^{-1})$ are, according to Equation (5.16) and the pair (5.52), (5.54), equal to one. By increasing the degree of polynomial $D(z^{-1})$ to 4-th degree, the degree of polynomials will then be, according to Equations (5.51)–(5.54), $\partial P(z^{-1}) = 2$ and $\partial Q(z^{-1}) = 1$, or $\partial P(z^{-1}) = 1$ and $\partial Q(z^{-1}) = 2$. The controller equation has one of the following forms

$$u(k) = R(z^{-1})w(k) - q_0 y(k) - q_1 y(k-1) - p_1 u(k-1) - p_2 u(k-2) \quad (5.55)$$

$$u(k) = R(z^{-1})w(k) - q_0 y(k) - q_1 y(k-1) - q_2 y(k-2) - p_1 u(k-1) \quad (5.56)$$

The courses of the control processes are equal for both types of controller. The design of controllers with polynomial $K(z^{-1})$ have only one solution according to Equations (5.53) and (5.54).

The parameters of the polynomials $P(z^{-1})$ a $Q(z^{-1})$ are equal for controllers with one and two degrees of freedom. The relationship for computing the coefficients of polynomial $R(z^{-1})$ for a controller with two degrees of freedom will now be derived. The ratio of polynomials according to (5.18) will be applied to the equation for tracking error (5.14) and the denominator of the transfer function will be replaced by $D(z^{-1})$ according to Equation (5.46) or (5.47)

$$E(z^{-1}) = W(z^{-1}) - Y(z^{-1}) = \left[\frac{D(z^{-1}) - B(z^{-1})R(z^{-1})}{D(z^{-1})} \right] \frac{N_w(z^{-1})}{D_w(z^{-1})} \quad (5.57)$$

The expression may be simplified if the polynomial $D_w(z^{-1})$ divides the expression in the numerator. Denoting this ratio as polynomial $S(z^{-1})$, it is possible to obtain the second conditional equation in the form

$$D_w(z^{-1})S(z^{-1}) + B(z^{-1})R(z^{-1}) = D(z^{-1}) \qquad (5.58)$$

For a step-changing value when $D_w(z^{-1}) = 1 - z^{-1}$, it is possible to solve Equation (5.58) by substituting $z = 1$

$$r_0 = \frac{D(1)}{B(1)} = \frac{1 + d_1 + \ldots + d_m}{b_1 + \ldots + b_n} \qquad (5.59)$$

If a controller contains the polynomial $K(z^{-1})$, the coefficient r_0 can also be computed, thus

$$r_0 = Q(1) = q_0 + q_1 + q_2 \qquad (5.60)$$

Note
The course of a control process is affected not only by the chosen poles, but also zeros in the numerator of the closed loop transfer function. Stable poles ensure the stability of a control process, but cannot always guarantee the required response properties – for instance, the size of overshoot or the time taken to settle. If a continuous-time transfer function of second order has only a constant in the numerator according to (5.39), then the numerator of the corresponding discrete transfer function contains a polynomial of the second-degree. The transfer function according to (5.4) has, for a controller with two degrees of freedom, the form

$$G_w(z) = \frac{Y(z)}{W(z)} = \frac{B(z^{-1})r_0}{D(z^{-1})}$$

since for a controller with one degree of freedom, it has the form

$$G_w(z) = \frac{Y(z)}{W(z)} = \frac{B(z^{-1})Q(z^{-1})}{D(z^{-1})}$$

It is obvious that the numerator of the first transfer function is closer to the required form than the numerator of the second one, although both forms are approximations. Major differences could lead to completely different flow of a control process.

A controller to control a second-order system without time delay with the transfer function

$$G_P(z) = \frac{B(z^{-1})}{A(z^{-1})} = \frac{b_1 z^{-1} + b_2 z^{-2}}{1 + a_1 z^{-1} + a_2 z^{-2}}$$

will be derived from Equation (5.47), which for four chosen poles has the form

$$(1 + a_1 z^{-1} + a_2 z^{-2})(1 - z^{-1})(p_0 + p_1 z^{-1}) +$$
$$(b_1 z^{-1} + b_2 z^{-2})(q_0 + q_1 z^{-1} + q_2 z^{-2}) = 1 + d_1 z^{-1} + d_2 z^{-2} + d_3 z^{-3} + d_4 z^{-4}$$

A system of linear equations can be obtained using the uncertain coefficients method $(p_0 = 1)$

$$\begin{bmatrix} b_1 & 0 & 0 & 1 \\ b_2 & b_1 & 0 & a_1 - 1 \\ 0 & b_2 & b_1 & a_2 - a_1 \\ 0 & 0 & b_2 & -a_2 \end{bmatrix} \begin{bmatrix} q_0 \\ q_1 \\ q_2 \\ p_1 \end{bmatrix} = \begin{bmatrix} d_1 + 1 - a_1 \\ d_2 + a_1 - a_2 \\ d_3 + a_2 \\ d_4 \end{bmatrix}$$

with its solution

$$p_1 = \frac{r_6}{r_1} \qquad q_0 = \frac{r_2 - r_3}{r_1} \qquad q_1 = -\frac{r_4 + r_5}{r_1} \qquad q_2 = \frac{d_4 + p_1 a_2}{b_2}$$

and with auxiliary variables

$$x_1 = d_1 + 1 - a_1 \qquad x_2 = d_2 + a_1 - a_2 \qquad x_3 = d_3 + a_2 \qquad x_4 = d_4$$

$$r_1 = (b_1 + b_2)(a_1 b_1 b_2 - a_2 b_1^2 - b_2^2) \qquad r_2 = x_1(b_1 + b_2)(a_1 b_2 - a_2 b_1)$$

$$r_3 = b_1^2 x_4 - b_2[b_1 x_3 - b_2(x_1 + x_2)] \qquad r_4 = a_1 [b_1^2 x_4 + b_2^2 x_1 - b_1 b_2(x_2 + x_3)]$$

$$r_5 = (b_1 + b_2)[a_2(b_1 x_2 - b_2 x_1) - b_1 x_4 + b_2 x_3] \quad r_6 = b_1(b_1^2 x_4 - b_1 b_2 x_3 + b_2^2 x_2) - b_2^3 x_1$$

Similar to the dead-beat method, the computed parameters are substituted into the equation of a controller with one degree of freedom

$$u(k) = q_0 e(k) + q_1 e(k - 1) + q_2 e(k - 2) + (1 - p_1)u(k - 1) + p_1 u(k - 2)$$

As mentioned above, the required control process is met by a controller with two degrees of freedom (denoted **PP1**) in the form

$$u(k) = r_0 w(k) - q_0 y(k) - q_1 y(k - 1) - q_2 y(k - 2) + (1 - p_1)u(k - 1) + p_1 u(k - 2)$$

where the parameter r_0 is computed from the relationship

$$r_0 = \frac{1 + d_1 + d_2 + d_3 + d_4}{b_1 + b_2}$$

If a smaller degree polynomial $D(z^{-1})$ is chosen (e.g. second order), it is sufficient to set coefficients d_3 and d_4 to zero. In the case of zero values for all coefficients d, the poles are placed at zero and the algorithm **PP1** is the same as algorithm **DB1**.

The algorithm **PP1**, applied to a control process with one degree of freedom according to Figure 5.2, is the same as algorithms PID–A1 and PID–A2 from Section 4.7.3, if the polynomial $P(z^{-1})$ according to Equation (4.118) is written as

$$P(z^{-1}) = (1 - z^{-1})(1 + \gamma z^{-1}) = K(z^{-1})(1 + p_1 z^{-1}) \qquad (5.61)$$

Also the algorithms PID–B1 and PID–B2, which use the structure designed by Ortega and Kelly [78] as shown in Figure 4.31, could, after formal adjustment of the defining polynomial, be replaced with the algorithm **PP1** in the following way: the transfer function of a closed loop described by Equation (4.136) is

$$G_W(z) = \frac{Y(z)}{W(z)} = \frac{\beta B(z^{-1})}{A(z^{-1})P(z^{-1}) + B(z^{-1})\left[Q'(z^{-1}) + \beta\right]} \qquad (5.62)$$

Polynomial $P(z^{-1})$ can be rewritten in the form (5.61) and polynomial $Q'(z^{-1}) + \beta$ can be defined as

$$Q'(z^{-1}) + \beta = (1 - z^{-1})(q_0' - q_2' z^{-1}) + \beta = q_0' + \beta - (q_0' + q_2')z^{-1} + q_2' z^{-2} \quad (5.63)$$

and can be replaced by the polynomial $Q(z^{-1})$ with coefficients $q_0 = q_0' + \beta, q_1 = -(q_0' + q_2')$ and $q_2 = q_2'$. Consequently, $\beta = r_0$. The advantage of the original method is that only four parameters are computed.

A different approach to controller design using pole assignment was suggested in [28]. The polynomial in the numerator of the transfer function $B(z^{-1})$ is factorized to

$$B(z^{-1}) = z^{-1}B^+(z^{-1})B^-(z^{-1}) \qquad (5.64)$$

and the controller polynomial $P(z^{-1})$ is chosen to be divisible by the stable part of the polynomial $B(z^{-1})$, thus

$$P(z^{-1}) = P_1(z^{-1})B^+(z^{-1}) \qquad (5.65)$$

Both denominator and numerator of the transfer function of closed loop (5.4) are then divided by the stable part of polynomial $B(z^{-1})$

$$
\begin{aligned}
G_w(z) &= \frac{B(z^{-1})R(z^{-1})}{A(z^{-1})K(z^{-1})P(z^{-1}) + B(z^{-1})Q(z^{-1})} \\
&= \frac{z^{-1}B^-(z^{-1})R(z^{-1})}{A(z^{-1})K(z^{-1})P_1(z^{-1}) + z^{-1}B^-(z^{-1})Q(z^{-1})}
\end{aligned}
\qquad (5.66)
$$

The conditional Equation (5.47) is then changed into the form

$$A(z^{-1})K(z^{-1})P_1(z^{-1}) + z^{-1}B^-(z^{-1})Q(z^{-1}) = D(z^{-1}) \qquad (5.67)$$

For computation of the controller parameters, the algorithm **PP1** can be used when the polynomial $B(z^{-1})$ is replaced with its unstable part.

5.4 Linear Quadratic Control Methods

Linear quadratic control (LQ) methods try to minimize the quadratic criterion by penalizing the controller output

$$J = \sum_{k=0}^{\infty} \left\{ [w(k) - y(k)]^2 + q_u[u(k)]^2 \right\} \tag{5.68}$$

where q_u is the so-called penalization constant, which gives weight of the controller output on the value of the criterion (where the constant at the first element of the criterion is considered equal to one). The standard procedure for minimization of criterion (5.68) is based on the state description of the system and leads to solution of the Riccati equation. In this section, criterion minimization is realized through spectral factorization for the input/output description of the system. LQ control methods, based on state variables, created by shifted inputs and outputs, are described in detail in Chapter 6.

If the sequences of the values of both tracking error and input signal are considered as polynomials, it is possible to rewrite criterion (5.68), using notation $\langle x \rangle = x(0)$

$$J = \left\langle E(z)E(z^{-1}) + q_u U(z)U(z^{-1}) \right\rangle \tag{5.69}$$

where $E(z)$ and $U(z)$ are the conjugated polynomials to the polynomials $E(z^{-1})$ and $U(z^{-1})$, which means their negative powers are replaced by positive ones.

The tracking error polynomial

$$\begin{aligned} E(z^{-1}) &= W(z^{-1}) - Y(z^{-1}) \\ &= \left[1 - \frac{B(z^{-1})R(z^{-1})}{A(z^{-1})K(z^{-1})P(z^{-1}) + B(z^{-1})Q(z^{-1})} \right] W(z^{-1}) \end{aligned} \tag{5.70}$$

and the input signal polynomial

$$U(z^{-1}) = \frac{A(z^{-1})R(z^{-1})}{A(z^{-1})K(z^{-1})P(z^{-1}) + B(z^{-1})Q(z^{-1})} W(z^{-1}) \tag{5.71}$$

are substituted into criterion (5.69) and the condition for criterion minimization is found. It can be verified [113] that the criterion is minimum if the equation

$$A(z^{-1})K(z^{-1})P(z^{-1}) + B(z^{-1})Q(z^{-1}) = D(z^{-1}) \tag{5.72}$$

is valid. The polynomial $D(z^{-1})$ is the result of spectral factorization according to the equation

$$A(z^{-1})q_u A(z) + B(z^{-1})B(z) = D(z^{-1})\delta D(z) \tag{5.73}$$

where δ is a constant chosen so that $d_0 = 1$. The spectral factorization of a polynomial leaves its stable part unchanged, while the unstable parts change to reciprocal ones (stable). Spectral factorization of polynomials of the first and the second degree can be computed simply; the procedure for higher degrees must be performed iteratively.

While performing the spectral factorization of a polynomial of the second degree $M(z^{-1})$, the following equation is solved

$$M(z^{-1})M(z) = D(z^{-1})\delta D(z) \qquad (5.74)$$

The products of the polynomials can be extended as

$$m_0 + m_1(z + z^{-1}) + m_2(z^2 + z^{-2}) = \delta(1 + d_1^2 + d_2^2) + \delta d_1(1 + d_2)(z + z^{-1})$$
$$+ \delta d_2(z^2 + z^{-2})$$

$$(5.75)$$

where the constants of the factorized polynomial on the left side of the equation are combined into the coefficients m_0, m_1 and m_2. Comparing the left and right side of Equation (5.75), one obtains

$$m_0 = \delta(1 + d_1^2 + d_2^2) \qquad (5.76)$$
$$m_1 = \delta d_1(1 + d_2) \qquad (5.77)$$
$$m_2 = \delta d_2 \qquad (5.78)$$

The conditions for stability of the polynomial $D(z^{-1})$ result from the Routh–Schur criterion

$$1 - d_2^2 > 0 \qquad (5.79)$$
$$(1 + d_2)^2 - d_1^2 > 0 \qquad (5.80)$$

Solving Equations (5.76)–(5.80), the following expressions are derived

$$\delta = \frac{\lambda + \sqrt{\lambda^2 - 4m_2^2}}{2} \qquad (5.81)$$

where

$$\lambda = \frac{m_0}{2} - m_2 + \sqrt{\left(\frac{m_0}{2} + m_2\right)^2 - m_1^2} \qquad (5.82)$$

$$d_1 = \frac{m_1}{\delta + m_2} \qquad (5.83)$$

$$d_2 = \frac{m_2}{\delta} \qquad (5.84)$$

As identical expression can be used for spectral factorization of polynomials of the first degree for $m_2 = 0$.

Solving the spectral factorization of Equation (5.73), an identical expression can be used, but it is necessary to convert the left side of this equation to the form used in Equation (5.74), thus

$$m_0 = q_u(1 + a_1^2 + a_2^2) + b_1^2 + b_2^2 \tag{5.85}$$

$$m_1 = q_u(a_1 + a_1 a_2) + b_1 b_2 \tag{5.86}$$

$$m_2 = q_u a_2 \tag{5.87}$$

The control algorithm based on the LQ control method (denoted **LQ1**), contains the following steps.

Step 1: The parameters of the polynomial $M(z^{-1})$ are computed according to Equations (5.85)–(5.87)

$$m_0 = q_u(1 + a_1^2 + a_2^2) + b_1^2 + b_2^2 \qquad m_1 = q_u(a_1 + a_1 a_2) + b_1 b_2 \qquad m_2 = q_u a_2$$

Step 2: The parameters of the polynomial $D(z^{-1})$ are computed according to Equations (5.81)–(5.87):

$$\lambda = \frac{m_0}{2} - m_2 + \sqrt{\left(\frac{m_0}{2} + m_2\right)^2 - m_1^2} \qquad \delta = \frac{\lambda + \sqrt{\lambda^2 - 4m_2^2}}{2}$$

$$d_1 = \frac{m_1}{\delta + m_2} \qquad d_2 = \frac{m_2}{\delta}$$

Step 3: The controller parameters are computed using the **PP1** algorithm. Penalization of the controller output is performed by setting $q_u \geq 0$. With increasing size of penalization constant, the amplitude of the controller output decreases and thereby, the flow of the process output is smoothed and any possible oscillations or instability are damped. For $q_u = 0$, the polynomial $D(z^{-1}) = B(z^{-1})$, and the method changes to the weak version of the deadbeat method. Thus, with variation of the penalization constant, it is possible to change the characteristics of the control loop, which will be demonstrated in the simulation examples in Section 5.5.

5.5 Simulation Experiments

Simulation of control with controllers, described in detail in previous chapters, has been realized with three systems of second order:
  A - stable;
  B - with nonminimum phase;
  C - unstable.
The continuous-time transfer functions of the systems (see also Example 4.1 in Chapter 4) are

$$G_A(s) = \frac{1}{(5s + 1)(10s + 1)} \qquad G_B(s) = \frac{1 - 4s}{(4s + 1)(10s + 1)}$$

$$G_C(s) = \frac{s + 1}{(2s - 1)(4s + 1)}$$

The discrete forms of the transfer functions for sampling period $T_0 = 2$ s are

$$G_A(z) = \frac{0.03286z^{-1} + 0.0269z^{-2}}{1 - 1.48905z^{-1} + 0.54881z^{-2}}$$

$$G_B(z) = \frac{-0.10166z^{-1} + 0.17299z^{-2}}{1 - 1.42526z^{-1} + 0.49659z^{-2}}$$

$$G_C(z) = \frac{0.6624z^{-1} + 0.01369z^{-2}}{1 - 3.3248z^{-1} + 1.64872z^{-2}}$$

The simulations were performed in the MATLAB® and SIMULINK® environments.

Because the main purpose of these simulations was to show the properties of the methods investigated, these algorithms are not adaptive. The speed of obtaining parameters during recursive identification is dependent on the initial state of the system, and it is assumed that the estimated parameters are at their nominal values because the systems do not contain disturbances. From the results of the Example 4.2, it follows that stabilization of the parameters is completed after approximately five sampling periods.

5.5.1 Dead-beat Methods

Example 5.2. Compute the sequences of the values of the process output and the **DB1** controller output for step changes of the reference for all three systems.

The output and input system variables have the following calculated values (see also Figure 5.5):

Stable system **A**:

$$y(0) = 0; \quad y(1) = 0.5498; \quad y(2) = 1; \quad y(3) = 1; \quad \cdots$$

$$u(0) = 16.7333; \quad u(1) = -8.1834; \quad u(2) = 1; \quad u(3) = 1; \quad \cdots$$

Nonminimum phase system **B**:

$$y(0) = 0; \quad y(1) = -1.4252; \quad y(2) = 1; \quad y(3) = 1; \quad \cdots$$

$$u(0) = 14.0193; \quad u(1) = -5.9619; \quad u(2) = 1; \quad u(3) = 1; \quad \cdots$$

Unstable system **C**:

$$y(0) = 0; \quad y(1) = 0.9798; \quad y(2) = 1; \quad y(3) = 1; \quad \cdots$$

$$u(0) = -1.4791; \quad u(1) = 3.4386; \quad u(2) = 1; \quad u(3) = 1; \quad \cdots$$

The **DB1** method ensures tracking of the step change of the reference from the second sampling period for all three systems, as follows from the theory.

Example 5.3. Compute the tracking of a reference signal in the form of a linear increasing function.

The transfer function of this reference signal can be written as the ratio of polynomials, thus

$$W(z^{-1}) = \frac{N_w(z^{-1})}{D_w(z^{-1})} = \frac{z^{-1}}{1 - 2z^{-1} + z^{-2}} \qquad (5.88)$$

The controller parameters, in comparison with Example 5.2, must be changed only in the polynomial $R(z^{-1})$ according to Equation (5.20)

$$r_0 = \frac{2 + r_1 b_2}{b_1} \qquad r_1 = \frac{-b_1 - 2b_2}{(b_1 + b_2)^2} \qquad (5.89)$$

The control process for system **A** is shown in Figure 5.4, where it is compared with the control process from Example 5.2, in which only step changes are expected. While in the first case, exact tracking of the reference can be observed from the third sampling period, the second controller leaves a permanent tracking error. Better tracking of the reference is achieved with greater values of the controller output (see right side of Figure 5.4). The control processes for the **B** and **C** systems have similar properties.

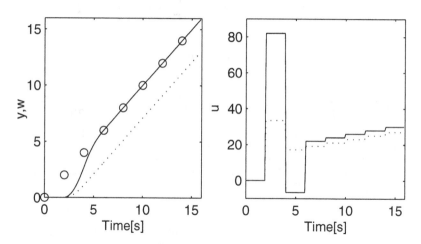

Figure 5.4. Control process for a linear increasing reference value (ramp) w – **DB1** method (solid line – controller with parameters according to (5.89); dashed line – controller designed for step changes in w; circles – discrete values of the reference w)

In conformity with the theory, the controller **DB1**, containing $K(z^{-1})$ ensures zero tracking error after the effects of the disturbance to the input of the system (denendent on the type of the system, maximally after four sampling periods).

Example 5.4. Repeat Example 5.2 for a controller with one degree of freedom, *i.e.* with only polynomials $P(z^{-1})$ and $Q(z^{-1})$.

This algorithm ensures that step changes are tracked from the fourth sampling period and results in a large overshoot (System **A**)

$$y(0) = 0; \quad y(1) = 1.7720; \quad y(2) = 1.5195; \quad y(3) = 0.6060; \quad y(4) = 1; \quad \ldots$$

Example 5.5. Compute the tracking of step changes in the reference for the weak version of the dead-beat method (**DB2**) for System **A** and compare with the **DB1** method.

Calculation results are demonstrated in Figure 5.5. Using the **DB2** method the tracking error is zero at the sampling times from the first sampling period, but between sampling times the process output oscillates. Compared to the **DB1** method, the control process is much worse.

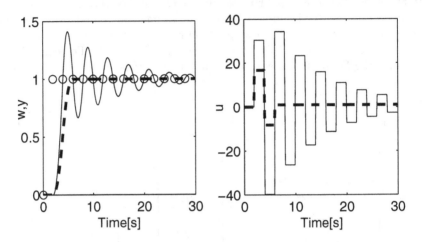

Figure 5.5. The control process for step change of the reference for System **A** (solid line – method **DB2**; dashed line – method **DB1**; circles – discrete values of the reference w)

The process control for System **C** is without oscillation, and stabilizes after three sampling periods. The **DB2** method leads, for the nonminimum phase System **B**, to an unstable control process. Because it is not possible to eliminate the creation of nonminimum phase models during recursive identification, the **DB2** method is not convenient for self-tuning controllers.

5.5.2 Pole Assignment Methods

Example 5.6. Design a **PP1** controller to ensure the same poles in the closed loop as the process in Example 5.1 in Section 5.3.

The denominator of continuous system (5.39) may be converted into discrete form using (5.45). The coefficients of the polynomial $D(z^{-1})$ are then $d_1 = -1.3413$ and $d_2 = 0.6188$. Process control with controller **PP1** is almost identical for Systems **A** and **C** with the required response according to Figure 5.3. For the nonminimum phase System **B**, the output variable does not track the chosen reference signal, but process control is acceptable (see Figure 5.6).

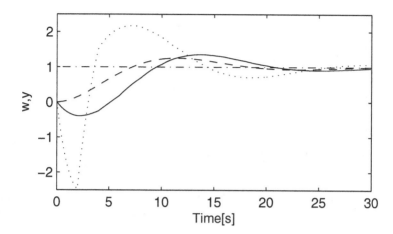

Figure 5.6. Process control for System **B** (dashed line – required response; solid line – control with controller **PP1**; dotted line – controller with one degree of freedom, dash and dot line – reference value w))

If a controller with one degree of freedom is used, the numerator of the closed loop transfer function contains the product of polynomials $B(z^{-1})$ and $Q(z^{-1})$, which causes an acceleration of the response but also increases overshoot and undershoot – see dotted line in Figure 5.6.

5.5.3 Linear Quadratic Control Methods

Example 5.7. Compute the response of control processes with the **LQ1** algorithm and differing sizes of penalization constant for System **B**.

The results are demonstrated in Figure 5.7. With increasing penalization constant, the response of the control process decelerates and undershoots decrease. Consequently, the values of controller output are reduced.

5.6 Summary of chapter

The most commonly used methods – dead-beat, pole assignment and LQ control, based on algebraic control theory, are presented in this chapter. For each method a general algorithm for controller design is derived, together with its concrete form for a second-order system. Properties of each method are shown by simulation examples. Close relations between methods are also mentioned. The methods described are given in basic form and their adaption into

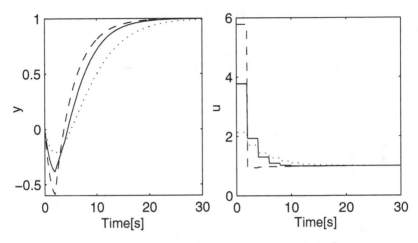

Figure 5.7. Control processes for the **LQ1** controller for System **B** and different penalization constants (dashed line $q_u = 0$; solid line $q_u = 0.03$; and dotted line $q_u = 0.2$)

self-tuning controllers involves repeated calculation of controller parameters according to the recursive identification model at every step.

Problems

5.1. Design a strong version of the DB controller for a first-order system with a time delay of two sampling periods. Compute the sequences of values of the process output and controller output for step and ramp changes of the reference signal.

5.2. Compute the sequence of values of the process output for a step disturbance present on the system input for the strong version of a DB controller with one degree of freedom, with and without polynomial $K(z^{-1})$.

5.3. Compare the sequences of the process output for controllers designed in line with Equations (5.37) and (5.38) during the control of nonminimum phase systems, given the transfer function $G_B(z)$.

5.4. Design the polynomial $D(z^{-1})$ for a **PP1** method which ensures a first overshoot of process output of about 30% in the time 10 s.

5.5. Compute the spectral factorization of the polynomial $M(z^{-1}) = 1 + 2.5z^{-1} + z^{-2}$. Observe the changes to its unstable part.

6

Self-tuning Linear Quadratic Controllers

In this chapter we will discuss the design of controllers using minimization of the quadratic criterion where the linear model of the system is known (the LQ method). In theory, the field of LQ control has seen detailed development. This chapter, however, concentrates mainly on the results of theoretical work and how they affect control behaviour in practice. This will be done by testing the results in simulations using the well-known MATLAB® SIMULINK® environment, which allows the user to experiment interactively. For the benefit of those who do not have access to MATLAB®, the SIMULINK® illustrations serve as block diagrams and the simulation results must be taken on trust.

The chapter is composed of several sections which may be of interest to a variety of readers.Section 6.1 and Section 6.2 provide a survey of the standard results of LQ control. There is a selection of examples to demonstrate the typical behaviour of LQ controlled loops and the effect of the basic "tuning" parameters representing the penalty of the control output in the quadratic criterion. This is used to achieve the desired control behaviour. In these examples we consider the match between the model and the system. Section 6.3 is devoted to the problems involved in using LQ designs in adaptive controllers. Section 6.4 gives an in-depth description of the characteristics of the LQ approach from various aspects and contains information that is readily available from textbooks, together with results from more hidden sources and from the work of the authors themselves. These are given here because our understanding of the individual aspects and properties of LQ control forms the basis for establishing a methodology to tune LQ controllers. This methodology is dealt with in Section 6.5. Section 6.6 briefly discusses extending the application of these methods to multivariable control loops. Section 6.7 deals with the algorithmic aspects of LQ synthesis and details the square root method of minimization of the quadratic criteria on which all versions of LQ controllers used here are based. There is also a reference to the related m-files for the MATLAB® environment.

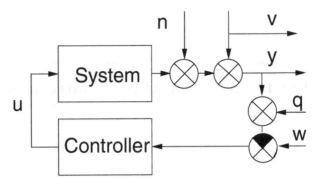

Figure 6.1. Diagram of a control loop

6.1 The Principles of Linear Quadratic Controller Design

We will take the basic control scheme pictured in Figure 6.1 as our starting point. This is a simplified and idealized version of reality. System is the process under consideration. This could, for example, be an electric oven. The temperature inside the oven at a given spot is the variable we wish to control, *i.e.* the output $y(t)$. The temperature depends on the energy supplied, which we usually vary by altering the electrical current, and on the temperature in surrounding area. The electrical current, therefore, is simultaneously system input variable $u(t)$ and the controller output. The effects of the other variables are represented by disturbance $n(t)$. In this example, disturbance is both the change in the voltage supply, which also influences power input, and changes in the surrounding temperature. Sometimes the effect of disturbance can be specified more precisely. The influence of the voltage supply can be defined with relative precision and, if we know how it behaves, it can be effectively compensated. We call this type of disturbance "measurable" and it is represented by signal $v(t)$ in the diagram. Output $y(t)$, which we measure for use in the control process, does not always correspond to the true physical variable. This difference is represented by signal $q(t)$. We will not take this signal into account but assume that we have a sufficiently precise measuring system. In the section dealing with the properties of LQ controllers we will show how its effects might be felt. The remaining variable $w(t)$ in the diagram represents the set point. In controller design it is necessary to know (select) the criterion and process of the model and to use a suitable optimization procedure. We will concern ourselves with the discrete approach where the signal is known at the instants of sampling only. The choice of sampling period is, therefore, an important factor in the design of an LQ controller for a continuous-time process. We will discuss the influence of the sampling period in Section 6.4, but for now we will assume that it was chosen in such a way as to provide a discrete model which describes the behaviour of the system well.

6.1.1 The Criterion

The aim of control is to generate the kind of input signal $u(k)$ which results in a system output close to the set point, *i.e.* so that

$$[w(k) - y(k)] \rightarrow min$$

for all the k values under observation. This means that the criterion used to evaluate the quality of control must be a non-negative function of all variables $(w(k) - y(k))$ at those values of k we wish to use in the design. These conditions are satisfied by the quadratic criterion

$$J = \sum_{i=k+1}^{k+T} [q_y(w(i) - y(i))^2] \tag{6.1}$$

When evaluating the quality of control achieved we must also consider the cost of a larger input signal to improve quality. This is reflected in the criterion by a further term

$$J = \sum_{i=k+1}^{k+T} [q_y(w(i) - y(i))^2 + q_u u(i)^2] \tag{6.2}$$

where q_y, q_u, T are the basic criterion parameters. We try to choose these in such a way as to make the criterion represent the user's definition of optimal control behaviour. Later we will discuss a further extension of the quadratic criterion to include a reference signal for the controller output. This criterion will have the form

$$J = \sum_{i=k+1}^{k+T} [q_y(w(i) - y(i))^2 + q_u(u(i) - u0(i))^2] \tag{6.3}$$

This makes the control output dependent on yet another purposely chosen signal, and we will show how this can be used to advantage in maintaining the useful properties of the controller..

6.1.2 The Model

The principle of LQ design is that it starts from current time k_0, and tries to generate values of $u(k_0 + 1)$, $u(k_0 + 2)$, ..., $u(k_0 + T)$, to keep the future error to a minimum. In order to be able to put the LQ method into practice, models must be available which allow one to determine the future values of all signals (variables) occurring in the criterion. In the simplest case this is a system model used to calculate the future values of output $y(k)$, and a model which determines the behaviour of the set point in the future. As far as the set point is concerned, it may be that its future numerical values are known rather than having to refer to the model.

One linear model which satisfies minimization of the quadratic criterion is the linear regression model given in Chapter 3. Because the model is a good one from the point of view of parameter identification and can be used in synthesis, it is the most frequently used in adaptive controllers. A regression model with measurable external disturbance (transformed into the system output) will be considered, *i.e.* regression model

$$y(k) = -\sum_{i=1}^{n} a_i y(k-i) + \sum_{i=0}^{n} b_i u(k-i) + e_k + \sum_{i=1}^{n} d_i v(k-i) + K \quad (6.4)$$

Note

We must bear in mind that a single difference equation (regression model) can represent several transfer functions. This is true in this example where model (6.4) represents the transfer functions shown in Figure 3.1.

Similarly the structure of a controller is represented by more transfer functions. Minimization of the criterion leads to a difference equation for optimal $u(k)$ in the form

$$u(k) = \sum_{i=1}^{n} S_i y(k-i) + \sum_{i=1}^{n} R_i u(k-i) + \sum_{i=1}^{n} Cld_i v(k-i) + \sum_{i=1}^{N} Clw_i w(k+i) \quad (6.5)$$

This corresponds to the diagram given in Figure 6.2.

We can see that the control law obtained simultaneously represents the feedback from the system output and the feedforward from the individually measurable variables. It follows that the controller here does not operate with a control error but independently with each individual signal. If we wish to compare the properties of the loop with a situation where control error is first introduced as controller input, the loop has to be transformed into the structure given in Figure 6.3.

The structure of the controller is closely bound to that of the model. All the transfer functions produced by the model are employed in the controller. If, for example, only the feedback part is used, a significant deterioration in the behaviour can be expected.

6.1.3 The Optimization Approach

The standard method of minimizing the quadratic criterion for LQ control is derived from the state description of the system. The state space formulation permits the elegant use of dynamic programming so that the minimizing process is performed from the end of the interval back to the beginning. In a linear state space model, step-by-step minimization results in the evolution of a matrix of quadratic form. This can be expressed in the form of an equation known as the (discrete) Riccati equation.

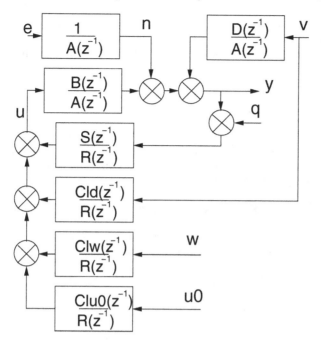

Figure 6.2. Diagram of a controller

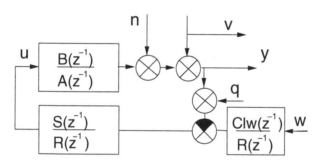

Figure 6.3. Diagram of the error approach

A Survey of Standard Results

In most texts linear quadratic control is presented in the following form. The
system is defined by the discrete state space model

$$\mathbf{x}(k) = \mathbf{F}\mathbf{x}(k-1) + \mathbf{G}\mathbf{u}(k-1)$$
$$y(k) = \mathbf{C}\mathbf{x}(k) + \mathbf{D}\mathbf{u}(k) \tag{6.6}$$

where vectors $\mathbf{x}(k)$, $\mathbf{y}(k)$, $\mathbf{u}(k)$ represent state, output and input vectors, and
have dimensions n, r, m. Matrices $\mathbf{F}$, $\mathbf{G}$, $\mathbf{C}$, $\mathbf{D}$ are the state, input, output

and feedforward matrices, and their dimensions correspond to those of the relevant vector.

The aim is to find the sequence of control laws $\mathbf{L}_i$ which minimize criterion

$$J = \sum_{k=1}^{T} \mathbf{x}(k)^T \mathbf{Q}_x \, \mathbf{x}(k) + \mathbf{u}(k)^T \mathbf{Q}_u \, \mathbf{u}(k) \tag{6.7}$$

i.e. optimize the transfer function of given state $\mathbf{x}(0)$ to zero state. $\mathbf{Q}_x$ and $\mathbf{Q}_u$ are penalty matrices for state and input, and we assume that $\mathbf{Q}_x \geq 0$ and $\mathbf{Q}_u > 0$ is valid. The discrete Riccati equation which represents the minimization takes the form

$$\mathbf{S}_i = \mathbf{F}^T \mathbf{S}_{i-1} \mathbf{F} - \mathbf{F}^T \mathbf{S}_{i-1} \mathbf{G} (\mathbf{Q}_u + \mathbf{G}^T \mathbf{S}_{i-1} \mathbf{G})^{-1} \mathbf{G}^T \mathbf{S}_{i-1} \mathbf{F} + \mathbf{Q}_x \tag{6.8}$$

where index i should be taken to be iteration $i \in [1, T]$, $\mathbf{S}_0 = 0$. The sequence of the matrices $\mathbf{S}_i$ defines the sequence of control laws

$$\mathbf{L}_i = (\mathbf{Q}_u + \mathbf{G}^T \mathbf{S}_{i-1} \mathbf{G})^{-1} \mathbf{G}^T \mathbf{S}_{i-1} \mathbf{F} \tag{6.9}$$

These control laws are applied in reverse order of iteration, *i.e.*,

$$\mathbf{L}_T, \ \mathbf{L}_{T-1}, .. , \mathbf{L}_2, \ \mathbf{L}_1$$

and optimal control is calculated from relation

$$\mathbf{u}(k + i) = \mathbf{L}_{T-i+1} \mathbf{x}(k + i - 1), \quad i = 1, 2, ..., T$$

The whole value of the criterion at the interval under consideration $[1, T]$ is given by the expression

$$J^* = \mathbf{x}(0)^T \mathbf{S}_T \mathbf{x}(0)$$

Using relation (6.9), the Riccati equation (6.8) can alternatively be written as

$$\mathbf{S}_i = \mathbf{F}^T \mathbf{S}_{i-1} (\mathbf{F} - \mathbf{G} \mathbf{L}_i) + \mathbf{Q}_x \tag{6.10}$$

$$\mathbf{S}_i = (\mathbf{F} - \mathbf{G} \mathbf{L}_i)^T \mathbf{S}_{i-1} (\mathbf{F} - \mathbf{G} \mathbf{L}_i) + \mathbf{Q}_x + \mathbf{L}_i^T \mathbf{Q}_u \mathbf{L}_i \tag{6.11}$$

The value of $T \to \infty$ has a special significance in control theory. In fact when we talk about the LQ problem we are almost always referring to this particular situation. Its importance lies in that the minimization of this criterion with an infinite horizon results in stationary control law $\mathbf{L}_\infty$, which is, therefore, time invariable. It has a stabilizing effect under standard conditions. These conditions affect the controllability (ability to stabilize) of pair $(\mathbf{F}, \mathbf{G})$ and the observability (detectability) of pair $(\mathbf{F}, \mathbf{Q}_x^{1/2})$. This control law is obtained from the solution of the algebraic Riccati equation. If it has several solutions we must choose the highest rank positively semidefinite solution.

This stationary solution has a number of other interesting properties which will be treated in another section. The most significant of these is the stabilizing effect, not only on the system it was designed for, but also on systems where the amplitude and phase frequency characteristics lie at a certain distance from the nominal. This kind of control is described as being robust in terms of stability. The difficulty is that the state must be accessible for the purposes of control. Access to state is rare in technological processes, so LQ optimum control will be formulated on the assumption that we only have access to the system output. The properties of the two types of control can differ greatly because another dynamic enters into the loop, either as a transfer function of a dynamic controller if it is relying on the system transfer function, or as the dynamics of an observer used to reconstruct an unmeasurable state. The guarantee of robustness and other qualities is lost if the state cannot be accessed. In our solution we use a state composed of delayed input and output. Although this state is made up of measurable signals, it is not itself measurable (to obtain these variables at time k, the previous measured values must be preserved, in shift registers perhaps, thus creating an observer). Therefore

$$\mathbf{x}(k-1) = [u(k-1), u(k-2), \ldots, u(k-n), y(k-1), y(k-2), \ldots, y(k-n)] \quad (6.12)$$

Here, the regressor of the model used is composed of the true input and the state vector

$$\mathbf{z}(k) = [u(k), \mathbf{x}(k-1)]$$

The state matrix of this type of state model is now made up of the parameters of the regression model; zeros and ones. The state used is not minimal. It can, however, be shown that the complexity of calculation involved here, especially in the square root form, is comparable to the complexity of the algorithm used to solve the minimal state problem.

Section 6.7 gives a detailed derivation of the minimization, including the square root algorithm. With regards to the state used, the calculated steady state control law contains the coefficients of polynomials R, S, and Clw Cld of the transfer function of controller (6.5).

Note

Must we really use the state formulation in the LQ problem if we are starting from the formulation of input/output and finally return to it? It is not entirely necessary but does carry certain advantages:

- The alternative, the polynomial approach, provides a solution only for the infinite horizon. However, in the adaptive approach, it is useful to imagine a steady solution as the limits of the finite horizon when the calculation method cannot find the steady solution.
- The state formulation allows us to formulate a cautious strategy and put it into practice – see the section on adaptive controllers.
- The nonminimum state approach permits the formulation of a strategy using a data-dependent penalty variable – see the section on adaptive controllers.

It is also possible to neglect the formulation of the state. An algorithm has been developed for the step-by-step minimization of the quadratic criterion which does not require the introduction of the state and state description. Unlike the classic approach, where the state is used to express the cost to go loss of the criterion using quadratic form $\mathbf{x}^T(k)\mathbf{S}_k\mathbf{x}(k)$, the new approach exploits a list of those variables contributing to the loss. This list is updated by individual steps of minimization. This update, however, can be far more generalized than merely relating to the state alone. This approach can be used on models where the structure changes along the criterion horizon, as well as to evaluate non-synchronous sampling for several variables [114, 115].The pseudo-state algorithms described here can be regarded as practicable due to their relative simplicity, transparency and reliability.

6.2 Using Linear Quadratic Controllers; Examples and Simulation

In the previous section we introduced the basic characteristics of controllers designed using the minimization of the quadratic criterion method. We are aware that we must select a criterion and a penalty for those variables included in the quadratic criterion, and determine the model which describes the behaviour of the controlled process. This is the system transfer function, the disturbance transfer function, and any other transfer functions from measurable variables. Having performed these two tasks, we use a computer program to generate the appropriate control law which can then represent also several transfer functions. It now remains to put the design into practice and discover what kind of control behaviour we have achieved. To make a good control law which meets our requirements, we must not only have a good process model, but we must also know how to select the penalty in the criterion which will lead to good control behaviour. The aim of this section is to acquaint the reader with the typical behaviour of an LQ controller working under different regimes and with various systems. The advantage of an LQ controller is that, mathematically speaking, the control behaviour is optimal with regard to the chosen criterion, at least when system and model match and are both linear. But not even these "ideal" conditions necessarily mean that behaviour is "user optimum". Better control behaviour almost always requires an input signal which consumes more energy, has a greater amplitude and a higher frequency. The user may be unable to permit such a signal. The user expresses the compromise between quality and the limitations on the input signal using penalty values on input Q_u (or more exactly the ratio Q_u/Q_y). This variable therefore becomes the basic tuning element. Shortly we will see that this tuning "button" is insufficient to adjust the control behaviour so that it meets the desired performance. In practice, other methods of modifying the control behaviour are required, such as dynamic penalties, modification of the transfer function of an open-loop or filtering. We will deal with these alternatives in a later section. We will familiarize ourselves with the typical characteristics of LQ controllers using examples from the MATLAB® and SIMULINK® environments and the functions and simulations in the LQ toolbox. This part will simultaneously introduce us to working with this toolbox and the possibilities it offers. The toolbox functions are fully listed and described in Section 7.2.

The control layout used for our simulation example features a system, controller, loops affected by disturbance, a set point generator and other auxiliary blocks. Each block must be defined before simulation starts; various procedures and programs are available in the toolbox to do this. The basic simulation procedure is represented in file *schema1.mdl* and is illustrated in Figure 6.4. This is similar to the scheme given in Figure 6.2.

Note

The diagram shows a system created by the *System* block using discrete transfer function and random disturbance generated by the *Discrete Noise* and *Disturbance* block. The *Saturation* block, which represents the limit on the controller output, is placed before the system. The controller is created by blocks *Filter–Filter 3.Filter* and *Filter 1* deal with feedback, *Filter 2* with the influence of the set point, and *Filter 3* with the effect of auxiliary signal $u0$, though this is not used in the standard approach and therefore disconnected from the diagram. The set point is generated by the SIMULINK® block *Signal Generator*. Other blocks are used to illustrate the input and output variables and calculate the sum of squares of the output error.

The *modely.m* file contains a number of pre-defined systems which can be selected interactively. The file also sets the corresponding sampling period T_0 and several other parameters in the SIMULINK® diagram, as well as the option of the controller calculation.

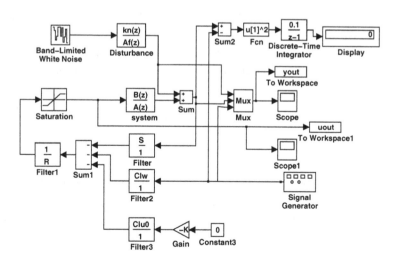

Figure 6.4. The basic simulation diagram

The quality of control behaviour depends to a large degree on the properties of the system. It is known that the so-called nonminimum phase system is among those in which the quality of control behaviour is sometimes severely limited. These are systems in which the transfer function has an unstable zero. While continuous-time nonminimum phase systems are not common, they occur more often with a discrete description. In this case the property of nonminimum phase does not only depend on the physical properties of the system but also on the sampling period.

Several simple systems have been chosen as examples. Nevertheless, they are well suited to demonstrating the chosen properties.

S1 A simple second-order system described by continuous-time transfer function

$$G(s) = \frac{1}{(s+1)^2} \tag{6.13}$$

In the main the discrete version is used, obtained by sampling at period $T_0 = 0.1\,\text{s}$, which has discrete transfer function

$$G(z^{-1}) = \frac{0.00468 + 0.00438z^{-1}}{1 - 1.81z^{-1} + 0.8187z^{-2}} \tag{6.14}$$

This is a simple process for achieving good quality control behaviour, it is minimum phase, and will be used to demonstrate standard behaviour.

S3 is one of the three discrete transfer functions representing one mechanical system introduced in [116] as a benchmark. The system transfer function, which has significantly elevated resonance, takes the form

$$G(z^{-1}) = \frac{0.2826z^{-3} + 0.50666z^{-4}}{1 - 1.9919z^{-1} + 2.2026z^{-2} - 1.8408z^{-3} + 0.8941z^{-4}} \tag{6.15}$$

S5 is a simple discrete model of a Pelton turbine with long feed pipes, where the transfer function is nonminimum phase and has the form

$$G(z^{-1}) = \frac{-2.0 + 2.05z^{-1}}{1 - 0.95z^{-1}} \tag{6.16}$$

The next four systems are constructed artificially. Discrete transfer function S1 was modified so that:

S6 the system has two time delay periods and so assumes the form

$$G\left(z^{-1}\right) = \frac{0.00468z^{-2} + 0.00438z^{-3}}{1 - 1.81z^{-1} + 0.8187z^{-2}} \tag{6.17}$$

S7 the system numerator contains one unstable root

$$G\left(z^{-1}\right) = \frac{0.1 - 0.16z^{-1} + 0.028z^{-2}}{1 - 1.81z^{-1} + 0.8187z^{-2}} \tag{6.18}$$

S8 , the system numerator contains two unstable roots

$$G\left(z^{-1}\right) = \frac{1 - 2.15z^{-1} + 1.155z^{-2}}{1 - 1.81z^{-1} + 0.8187z^{-2}} \tag{6.19}$$

S9 , the system is unstable and has the numerator from system S1

$$G\left(z^{-1}\right) = \frac{0.00468 + 0.00438z^{-1}}{1 - 2.104z^{-1} + 1.1067z^{-2}} \tag{6.20}$$

- Helicopter model. The SIMULINK® scheme of a simple helicopter model has been created. It represents the dynamics in elevation angle. This continuõus-time nonlinear model serves to demonstrate the features of an adaptive LQ controller used with a nonlinear system.

The basic optimization procedure *lqex1.m* from the toolbox is used in the optimal control calculation. The examples are divided into disturbance compensation and set point control to demonstrate the properties of a step response. The systems best suited to demonstrating these properties were selected from those above and used in the following simulations.

6.2.1 Stochastic Disturbance Compensation

The optimal control law will be composed purely of feedback component $\frac{S}{R}$. The compensation behaviour and the controller output for "well-behaved" system S1 and for two penalty values Q_u is captured in Figure 6.5. The graphs show the time behaviour of the output and disturbance without compensation (Sx-y, v) and controller output (Sx-u).

The next four graphs (Figure 6.6) show the error of disturbance compensation for oscillating system (S3), time delay system (S6), nonminimum phase system (S7), and unstable system (S9), when the penalty is relatively small $Q_u = 0.001$.

Note

1. The disturbance behaviour of systems S1, S6, S7, and S9 are the same because they were created by the same filter. In order to obtain a stable signal, the roots of the filter for unstable system S9 are reciprocal to its unstable roots. Its auto-correlation function, however, also corresponds to the denominator of the original unstable system.
2. If we want to experiment with the control of system S1 by changing the penalties, introducing an effective limitation on the controller output in the saturation block, or by using just a short horizon in the optimization criterion, most of the results will be obtained when the disturbance is more or less compensated for. At the same time, penalty Q_u greatly affects the quality of the disturbance compensation.

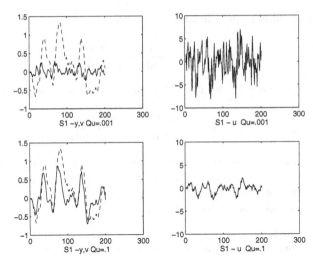

Figure 6.5. Disturbance compensation for system S1 with various penalties

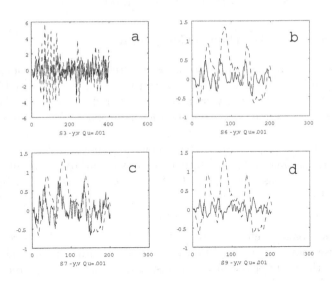

Figure 6.6. Disturbance compensation for different systems and with penalty $Q_u = 0.001$

3. Nothing like this, of course, applies to the other systems, which are either unstable, nonminimum phase or have time delay. For unstable system S9 the limitation of the input signal is critical. In the simulation shown in Figure 6.6d the controller output remains mostly within the boundary $|u(k)| < 5$. Nevertheless, applying this limitation results in a noticeable

deterioration in compensation, yet limitation $|u(k)| < 4$ leads to instability. It helps to increase penalty Q_u so that the limitation cannot come into effect. The length of the criterion horizon used in this case makes little difference to the quality of compensation.

4. The disturbance compensation for the nonminimum phase system S7 Figure 6.6c is poor. Further experiments show that it is almost independent of penalty Q_u. The amplitude of the controller output is small and its limitation by the saturation block results in a further deterioration in disturbance compensation, but in such a way as to approach disturbance without control. Here, the length of the horizon in the criterion used is significant. If a short horizon is used ($hor = 3$), the control behaviour is worse than if there were no control at all. Unless we consider the use of an extra penalty on the finite state, behaviour will be unstable (see Section 6.4.1 – Stability).

5. Systems S6 and S1 differ in that S6 has two time delay intervals. Compensation is similar to that in S1 and is illustrated in Figure 6.6b. Even when the horizon used in this system is short there is a penalty which can be applied to produce good disturbance compensation.

6. Figure 6.7 shows two types of disturbance compensation for system S5. The upper series shows the behaviour resulting from a single step strategy ($hor = 1$). A noticeable disturbance compensation can be seen but there is an unstable input signal. This is a typical feature in the control of a nonminimum phase system where satisfactory short-term control behaviour can be achieved with an unstable input signal. Optimal, stable behaviour can be seen in the lower pair of illustrations. The noise compensation is minor but requires just a very small input signal.

The compensation of disturbance represented by a regression model is truly optimal. In a minimum phase system, disturbance can be compensated to its source, white noise, which will remain in the compensation process. Compensation in nonminimum phase systems is limited. The advantage of an optimal controller here is that only a small controller output is required.

6.2.2 Set point Control

First a step set point change is considered. Figure 6.8 shows the step responses for system S1 and controller outputs using two different penalties. A similar situation is pictured in Figure 6.9, this time for nonminimum phase system S7.

Figure 6.10 illustrates the control behaviour of system S3 and unstable system S9. Figure 6.11 shows the control behaviour of systems S3 and S1 with a limited control output. Although S3 shows a poor result, limitation u has a positive effect on system S1. Experimenting with these systems under various conditions yields the following conclusions:

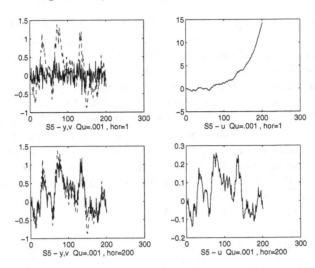

Figure 6.7. The disturbance compensation behaviour of system S5 for various criterion horizons

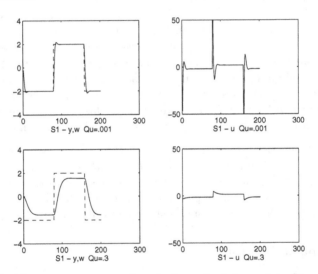

Figure 6.8. The step response and controller output for system S1 using various penalties

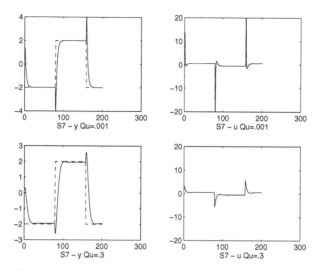

Figure 6.9. The step response and controller output for system S7 using various penalties

1. Standard optimization of the criterion results in a steady state control error. The size of this depends on the controller output penalty, and system gain. This is why penalty $Q_u = 0.001$ is not noticeable. An integrator term must be included in the open-loop or signal $u0$ must be employed to eliminate this error.
2. There are systems, S3 for example, where overshoot in the step response cannot be eliminated by any type of penalty (see Figure 6.10a).
3. As with disturbance compensation, limitation of the controller output is acceptable for a minimum phase system. However, it can cause instability in an unstable system and have a very negative effect in a nonminimum phase system.
4. It can be seen that the conditions for good control behaviour under set point changes are the same as for disturbance compensation.
5. The steady state control error may increase ramp-like, if the set point change is also ramp-like (Figure 6.12).

In the examples so far, the simulation of the systems has been discrete. It remains to test if the controller under consideration behaves in the same way in a continuous-time system. To do this we use simulation diagram *schema2.mdl*, where the system is simulated in continuous-time form. Figure 6.13 illustrates combined control for step response and disturbance compensation for systems S1 and S1 with time delay (in a discrete representation of S6). Since a suitable sampling period was used, all have the same type of behaviour.

In this section we have shown the typical behaviour of LQ controllers on selected examples. All behaviour was optimal, but we did not always obtain

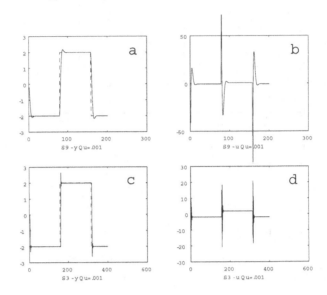

Figure 6.10. The step response and controller output for systems S3 and S9 for $Q_u = 0.001$

the type of behaviour required by the user. The main shortcomings were identified as being:

- a steady state control error
- oscillation in the transfer function process
- unstable behaviour when the criterion horizon is short
- instability when the system input signal is limited.

The next section gives a more detailed analysis of the solution to these problems, the questions raised by the sampling period, the issue of robustness, and general control properties where the controller was originally developed for a different system.

6.3 Adaptive Control

Sections 6.1 and 6.2 devoted to an introduction to the typical behaviour of LQ controllers. In all cases the system was known. In many situations the system is not known, is known only partially, or the plant is so complex that even if it is known the model used in the controller design can only be an approximation of the actual plant. In all these cases the first step in controller design is construction of a model. The structure of the model and its parameters are determined from information about the system (prior information) and from measured data on the process. It is obvious, however, that such a model is not

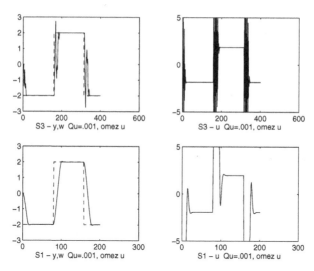

Figure 6.11. The step response and controller output for systems S1 and S3 when the controller output is limited

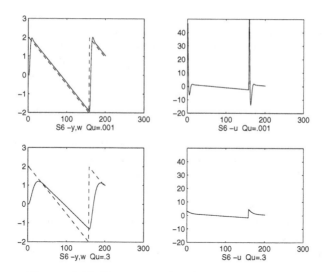

Figure 6.12. Ramp set point change response of S6

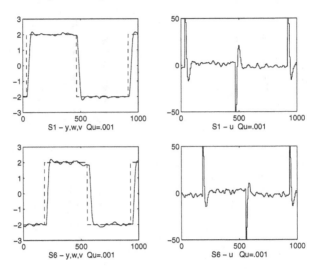

Figure 6.13. The response behaviour of continuous-time systems S1 and S6 to a step set point change and noise compensation

equivalent to reality. The behaviour of a controller designed to the model but used with an actual plant depends not only on the criterion and model used but also how the model represents the plant. In such situations adaptation and robustness start to become important. The aim of adaptivity is to tune the model in such a way that it tries to be, in some respect, close to reality. Robustness implies trying to guarantee that some control properties are maintained to be kept for all systems from supposed set. These two branches of control theory are often considered in isolation but in practice they are both needed simultaneously. Any modelling of a control process is limited to restricted-complexity models. The information about its parameters is also limited. The more the model is able to adapt to reality the less robustness is necessary, and vice versa.

It is clear that when adaptation takes precedence we achieve better control behaviour, so this section will be devoted to this topic. Identification plays a dominant role in adaptation. This is the process by which we learn about the properties of the system from measured data. The data can, for the time being, be regarded as random processes. We will first try to formulate the task of LQ control as a stochastic example. We will demonstrate how the identification process and control synthesis fit into the adaptive approach and the resulting specification for these processes. In conclusion we will again show the typical behaviour of an LQ controller using the LQ toolbox.

6.3.1 The Stochastic Approach to Linear Quadratic Controller Design

So far we have examined the characteristics of LQ design with regard to deterministic signals. LQ synthesis, however, lends itself to working with random signals. We need only recall the Wiener synthesis, which made it possible to design optimal control that minimized quadratic loss for a familiar stochastic disturbance.

If the system has random output or contains a random component, criterion (6.2) cannot be directly minimized. Instead we must use a deterministic function of random variables. Experience shows that the mean value of a random variable or process is an acceptable function. The criterion appears as follows

$$J = \mathcal{E} \left\{ \sum_{i=k+1}^{k+T} \left[q_y \left(w\left(i\right) - y\left(i\right) \right)^2 + q_u u\left(i\right)^2 \right] \right\} \qquad (6.21)$$

The introduction of mean value $\mathcal{E}$ does not involve any great complications. To minimize the criterion a model which defines the required mean value will be needed. This is regression model (6.4), which has already been introduced, where the deterministic part of the signal

$$\hat{y}\left(k\right) = - \sum_{i=1}^{n} a_i y(k-i) + \sum_{i=0}^{n} b_i u(k-i) + \sum_{i=1}^{n} d_i v(t-i) + K \qquad (6.22)$$

is simultaneously the mean value of output $y(k)$. This can be written as

$$\hat{y}(k) = \mathcal{E} \left\{ y\left(k\right) | \mathbf{x}\left(k-1\right), u\left(k\right) \right\} \qquad (6.23)$$

where $\mathbf{x}(k-1)$ contains the previous inputs and outputs, and signal $y(k)$ is modelled as

$$y(k) = \hat{y}(k) + e_s(k) \qquad (6.24)$$

Since $e_s(k)$ is white noise with zero mean value, $\mathcal{E}\left\{e_s(k)\right\} = 0$, the terms with $e_s(k)$ do not appear in the criterion. However, because $\mathcal{E}\left\{e_s^2(k)\right\} = \sigma_{e_s}^2$, the value of the criterion will increase with each new mean value of σ_{e_s}. This increment does not depend on $u(k)$ and cannot be influenced in any way. It can therefore be ignored when the criterion is minimized.

All this leads to the "Certainty Equivalence Principle", which states that when we do not know the true parameters we use the mean value of their estimates. These mean values are not necessarily acceptable as parameters. The start of adaptation is a typical situation where the parameter estimates are wildly off the correct values.

When this principle is used, synthesis is not based on information on the state (precision) of the estimated parameters contained in the covariance matrix. Attempts have been made to modify the synthesis so as to give a role to the quality of identification. Because the resulting control algorithm takes account of the identification state, this control strategy is referred to as being "*cautious*". Reference [117] is an example of this. Similar results can be obtained from generalizing the regression model so that the dispersion variance of the error prediction is a function of the covariance matrix of the parameters. This model has been given the name linear stochastic transform [118]. Whereas Equation (6.24) yields $\sigma_y^2 = \sigma_{e_s}^2$ for the regression model, the stochastic linear transform uses

$$\sigma_y^2 = \sigma_{e_s}^2 (1 + \xi) \tag{6.25}$$

where $\xi = \mathbf{z}^T(k)\mathbf{C}(k)\mathbf{z}(k)$, and $\mathbf{z}(k)$ is the data used in prediction $y(k)$. Here it is seen that noise dispersion is dependent on the value of $u(k)$, which is a part of $\mathbf{z}(k)$.

This model forms the basis for the so-called cautious strategy [118, 25], Despite partial success in practical applications, it is not often used in adaptive control due to two serious failings:

- The start-up covariance matrix $\mathbf{C}$ must be chosen very carefully if the strategy is to yield good results.
- The strategy very often chooses $u(k)$ to minimize ξ, and not its own criterion. The sequence of $u(k)$ is particularly ill-suited to the identification of parameters, with the result that the covariance matrix does not grow smaller. This means that the entire process cannot avoid bad control.

6.3.2 The Synthesis of Quadratic Control in Real Time

The use of LQ synthesis in adaptive controllers is governed by the fact that synthesis of the controller must be constantly repeated as the parameters alter. Generally speaking therefore, parameter estimation and synthesis are updated at each sampling period. The restriction placed on the final permissible calculation time is also important. On the one hand this is dependent on the sampling period and speed of the calculation technology, and on the other it depends on the complexity of the calculation. A calculation method which requires a constant and, if possible, short calculation time is best suited to adaptive control. Since this is an iterative process the in which calculation time depends on a number of factors, there is no guarantee that optimal control corresponding to the infinite horizon can be calculated in one control period. There are two alternatives:

(a)In each sampling period a finite horizon criterion will be minimized. The length of the horizon must allow enough space for the iterations to complete the Riccati equation in the time available. A test on convergence of the control law may be added to this approach and the calculation aborted if there is little change. Where the iterations are few in number, the start-up conditions for solving the Riccati equation play a vital role in stability and quality (the roots of a closed loop) – see Section 6.4.1. This is the matrix $\mathbf{S_0}$ which acts as the tuning "button". This approach, in which the horizon is shifted further and further away during the control process, is known as the *receding horizon strategy* or *moving horizon strategy* MH.

(b) If we require control with an infinite horizon we must spread the iterations of the Riccati equation in time. This means that a fixed number of iterations is performed in each control period, based on the previously attained state, and not on the start-up conditions. The effect of doing this is that the criterion horizon increases by the chosen number at each control period (NSTEP). After a certain time the control law will approach a steady state solution. This assumes, of course, that we use the same model parameters for the calculation in each control period. This is not necessarily true in adaptive control. It is impossible to tell what the control law will be when the parameters are apt to change. Experience shows that this strategy, known as *iteration spread in time* (IST), yields good results even when just one iteration is used during the control period, this being the shortest possible calculation time.

The IST strategy has another positive effect. It is both a one-step and a stabilizing strategy. When a strategy has just one step, the future input and output values are easy to determine and compare with the restriction requirement, so, for example, penalty Q_u can be directly modified to maintain the restriction. This algorithm is known as the MIST (modified iteration spread in time), and is described in [119] and [120] and later in Section 6.5.1.

6.4 The Properties of a Control Loop Containing an Linear Quadratic Controller

Before exploiting the possibilities offered by controllers based on the optimization of the quadratic criterion we must be more familiar with their properties. These can be considered from several angles. We can start with stability. The issue of stability in the control loop using a given controller is vital: if behaviour becomes unstable none of the other properties are relevant. The stability requirement is met by the classic PID type controller and is often used as the basis for design. A main factor here is the gain of the open-loop (see Ziegler–Nichols), and, on further analysis, also the gain at various frequencies, *i.e.* the frequency characteristic. The difficulties in maintaining stability when using an LQ controller, however, are very different.

We will then consider the properties of a control loop using an LQ controller in the time domain. This is vital in LQ control because its design is based on minimization of the criterion, which itself evaluates the time behaviour of the signals.

In the time domain the behaviour of the loop in the two usual cases, disturbance compensation and set point change, will be observed.

Although in a linear system there is an unambiguous relationship between the frequency and time domain characteristic, there are a number of properties which are better observed from the aspect of frequency. Robustness is a major example of these, since by this we mean observing the stability and quality of the control behaviour in cases where the model featured in the controller design differs from the true system.

6.4.1 Stability

Stability is probably the most important requirement of a control loop. The advantage of LQ design is that the stability requirement is, to a certain extent, automatically built into the controller design. This statement must include the qualifier "to a certain extent", because stability can only be theoretically guaranteed if the model used is a precise match to reality. In practice, the problem of stability is more complicated. Stability must be ensured even when reality differs from the mathematical model. This issue will be dealt with in the section on robustness of design.

The theoretical problems with stability have been published in several articles and there is a summary of conclusions in, for example, reference [121], the basic results of which we will give here. The minimization of quadratic criterion (6.2) results in the solution of the Riccati equation (6.8), the properties of which can then be used to manipulate stability. Stability is only naturally guaranteed for infinite horizon $T \to \infty$, where the solution is given by the so-called algebraic Riccati equation (ARE)

$$\mathbf{S} = \mathbf{F}^T \mathbf{S} \mathbf{F} - \mathbf{F}^T \mathbf{S} \mathbf{G} (\mathbf{G}^T \mathbf{S} \mathbf{G} + \mathbf{Q}_u)^{-1} \mathbf{G}^T \mathbf{S} \mathbf{F} + \mathbf{Q}_x \qquad (6.26)$$

Theorem 1. We will consider (6.26) corresponding to minimization (6.2) for $T \to \infty$ where

- $[\mathbf{F}, \mathbf{G}]$ is stabilizable
- $[\mathbf{F}, \mathbf{Q}_x^{1/2}]$ has no observeable mode on the unit circle
- $\mathbf{Q}_x \geq 0$, $\mathbf{Q}_u > 0$

  then

- there is a single, maximum rank, non-negative definite, symmetric solution $\bar{\mathbf{S}}$
- $\bar{\mathbf{S}}$ is the only stabilizing solution, and the matrix of closed loop $\mathbf{F} - \mathbf{G}(\mathbf{G}^T \bar{\mathbf{S}} \mathbf{G} + \mathbf{Q}_u)^{-1} \mathbf{G}^T \bar{\mathbf{S}} \mathbf{F}$ has eigenvalues inside the unit circle.

The matrix relation type $\mathbf{A} > \mathbf{B}$ can be seen as the definiteness of the matrix. Therefore matrix $\mathbf{A}$ is greater than $\mathbf{B}$ ($\mathbf{A} > \mathbf{B}$), provided that $\mathbf{A} - \mathbf{B} > 0$, so $\mathbf{A} - \mathbf{B}$ is a positively definite matrix.

Stability where the horizon is finite can easily be solved by transferring it to an infinite horizon using the following trick. Let the solution to the Riccati equation corresponding to the horizon of criterion (6.2) T be S_T. This simultaneously relates to ARE solution (6.26) for other state penalties $\bar{\mathbf{Q}}_x$.

$$\mathbf{S}_T = \mathbf{F}^T \mathbf{S}_T \mathbf{F} - \mathbf{F}^T \mathbf{S} \mathbf{G} (\mathbf{G}^T \mathbf{S}_T \mathbf{G} + \mathbf{Q}_u)^{-1} \mathbf{G}^T \mathbf{S}_T \mathbf{F} + \bar{\mathbf{Q}}_x \qquad (6.27)$$

where $\bar{\mathbf{Q}}_x = \mathbf{Q}_x - (\mathbf{S}_{T+1} - \mathbf{S}_T)$. Equation (6.27) is called the fake algebraic Riccati equation (FARE).

Stability when minimization of the criterion is used is then solved by the following theorem:

Theorem 2. We will consider Equation (6.27) which defines $\bar{\mathbf{Q}}_x$. If $\bar{\mathbf{Q}}_x \geq 0$, $\mathbf{Q}_u > 0$, $[\mathbf{F}, \mathbf{G}]$ is stabilizable, and $[\mathbf{F}, \mathbf{Q}x^{1/2}]$ is detectable, then $\mathbf{S}_T$ is stabilizing, and $\mathbf{F} - \mathbf{G}(\mathbf{G}^T\mathbf{S}_T\mathbf{G} + \mathbf{Q}_u)^{-1}\mathbf{G}^T\mathbf{S}_T\mathbf{F}$ has eigenvalues inside the unit circle.

It is clear from the definition of $\bar{\mathbf{Q}}_x$ that, if the sequence of $\mathbf{S}_T$ decreases, the stability requirement will automatically be satisfied since $\mathbf{S}_{T+1} - \mathbf{S}_T$ will be negatively semi-definite. There is a very strong relationship between monotonicity and stability in the Riccati equation. This can be used as the basis for formulating conditions which guarantee that the solution to the Riccati equation will have a stabilizing effect from a given iteration $k < T$. $k = 1$, i.e. the very start, is a special situation.

The theoretical basis creates a theorem on the monotonicity of the solution to the Riccati equation and the relationship between monotonicity and stability [122]. The following theorem evaluates the monotonicity of the Riccati equation:

Theorem on monotonicity: If the Riccati equation has a non-negative solution in iterations i, $i + 1$, $\mathbf{S}_i$, $\mathbf{S}_{i+1}$ and if $\mathbf{S}_i > \mathbf{S}_{i+1}$, is true, then $\mathbf{S}_{k+i} > \mathbf{S}_{k+i+1}$, is valid for all $k > 0$.

Theorem on stability (a): We consider difference the Riccati Equation (6.8). If

- $[\mathbf{F}, \mathbf{G}]$ is stabilizable
- $[\mathbf{F}, \mathbf{Q}_x^{1/2}]$ is detectable
- $\mathbf{S}_{i+1} \leq \mathbf{S}_i$, for some i

then the closed loop defined by $(\mathbf{F} - \mathbf{G}(\mathbf{Q}_u + \mathbf{G}^T\mathbf{S}_k\mathbf{G})^{-1}\mathbf{G}^T\mathbf{S}_k\mathbf{F})$ is stable for all $k \geq i$.

Similarly:

Theorem on stability (b): We consider the difference Riccati Equation (6.8). If

- $[\mathbf{F}, \mathbf{G}]$ is stabilizable
- $[\mathbf{F}, \mathbf{Q}_x^{1/2}]$ is detectable
- $\mathbf{S}_{i+2} - 2\mathbf{S}_{i+1} + \mathbf{S}_i \leq 0$, for some i

then the closed loop defined by $(\mathbf{F} - \mathbf{G}(\mathbf{Q}_u + \mathbf{G}^T\mathbf{S}_k\mathbf{G})^{-1}\mathbf{G}^T\mathbf{S}_k\mathbf{F})$ is stable for all $k \geq i$.

Before discussing the practical significance of these theorems, we should recall the physical effect of the individual iterations of the Riccati equation in the development of losses. These properties are most obvious when the aim of control is to transfer the system from initial conditions $\mathbf{x}_0$ to zero ($\mathbf{x}_\infty = 0$). Here the Riccati matrix (the R.E. solution) represents the development of a matrix of quadratic form determining the loss (the value of the criterion). If we have constant $\mathbf{Q}_u$, $\mathbf{Q}_y$, the loss will grow from the initial value of $\mathbf{S}_0$ to value $\mathbf{S}_T$. If at the time $k_0 + T$, the value $y(k_0 + T)$ is very small and consequently $u(k_0 + T)$ is also small, there will not be a great difference between $\mathbf{S}_T$ and $\mathbf{S}_\infty$.

If we take the iterations from the initial condition $\mathbf{S}_0 = 0$, corresponding to the formulation of criterion (6.2), the quadratic form must increase and so we never have the situation where $\mathbf{S}_i > \mathbf{S}_{i+1}$.

If we want to ensure stability we must find $\mathbf{S}_0$, $\mathbf{Q}_y$, $\mathbf{Q}_u$ such that $\mathbf{S}_0 > S_1$, or $\mathbf{S}_{i+2} - 2\mathbf{S}_{i+1} + \mathbf{S}_i \leq 0$. This is not possible with basic criterion (6.2) where $\mathbf{S}_0 = 0$ is supposed. $\mathbf{S}$ must increase in some way as it results from the principle of cumulative loss which $\mathbf{S}$ represents.

When stability is to be ensured using the quadratic criterion with a finite horizon, the criterion must be changed to include a penalty on the final state represented by matrix $\mathbf{S}_0$. The new criterion takes the form

$$J = \sum_{i=k+1}^{k+T} \left[q_y \left(w\left(i \right) - y\left(i \right) \right)^2 + q_u u\left(i \right)^2 \right] + \mathbf{x}^T(T)\mathbf{S}_0\mathbf{x}(T) \qquad (6.28)$$

$\mathbf{S}_0 > \mathbf{S}_\infty$ must be true if the sequence of $\mathbf{S}_i$ is to decrease monotonically. The penalty on the final state $\mathbf{S}_0$ must be greater than the entire cumulative loss of the whole compensation process of the initial state. Then cumulative loss may decrease, since the loss of each individual step in the minimization process is compensated for by a decrease in the value of the final state, and therefore a decrease in its contribution to the overall loss. The following example explains how it is possible for the cumulative loss to decrease. If we consider 0-step control, loss is $J_0 = \mathbf{x}_0^T \mathbf{S}_0 \mathbf{x}_0$. In one-step control, the overall loss can be written as $J_1 = \mathbf{x}_1^T \mathbf{S}_0 \mathbf{x}_1 + \mathbf{x}_0^T \mathbf{Q}_x \mathbf{x}_0$ (the finite state is now $\mathbf{x}_1$), and this value may be less than J_0 because value $\mathbf{x}_1$ is already smaller than $\mathbf{x}_0$ due to control. This process may be continued.

As stated in reference [121] and as can be seen in the example, not even the choice $\mathbf{S}_0 > \mathbf{S}_\infty$ can guarantee that $\mathbf{S}_0 > \mathbf{S}_1$. This also applies to the simple choice $\mathbf{S}_0 = \alpha I$, where α is a real positive number with a high value. No high value of α can ensure stability in feedback from the very beginning of the solution of the Riccati equation. One way of finding $\mathbf{S}_0$ to satisfy $\mathbf{S}_0 > \mathbf{S}_1$ is given in reference [123]. However, this involves inversion of the system matrix $\mathbf{F}^{-1}$. This cannot be applied to our example because the state matrix we use is always singular.

Initial value $\mathbf{S}_0 \neq 0$ means an extra penalty on the finite state. If T is large, it will have only a small effect on the quality of control. However, if the horizon is small, the choice of $\mathbf{S}_0 \neq 0$ can significantly alter control behaviour.

These remarks lead to the conclusion that it is relatively difficult to guarantee stability in a closed loop for a finite horizon T. Still, the situation is not entirely hopeless.

1. The stability requirements described above are sufficient, but not necessary; the examples given here will show these conditions are very strict.

2. Stability need not be guaranteed right from the very first iteration. An analysis of selected examples will guide us in our choice of horizon length.

3. We will look at cases where we can exploit knowledge previously obtained on stabilizing control or *a priori* knowledge of **S** gained from the minimization of a similar criterion (perhaps for a different value of Q_u or T, *etc*).

In order to ensure the stability of an LQ controller it is wise to:

- select a sufficiently long horizon for an off-line calculation and then check that some of the sufficient stability conditions are met;
- use the IST strategy in an adaptive environment as this will help us achieve an asymptotic infinite horizon.

The Examples

The evolution of the Riccati equation iteration understandably differs from system to system, and from penalty to penalty. We will use our chosen examples to demonstrate at least a few typical situations. The following illustrations show the position of the roots of a closed loop which would be obtained using a controller derived from the ith iteration of the Riccati equation. They will also illustrate the first and second differences of this equation (or more precisely, the eigenvalues of the matrix of the first and second difference of the Riccati equation for individual iterations $i = 1, 2, 3, ..., T$.

We can influence the number of iterations of the Riccati equation required to achieve stability using penalty Q_u and the penalty on the finite state $\mathbf{S}_0$. We will take $\mathbf{S}_0 = \alpha I$. This will be demonstrated on stable minimum phase system S1, unstable system S6 and nonminimum phase system S8.

Figures 6.14 and 6.15 show examples of a stable minimum phase system. Here we can see that stabilizing control can be achieved regardless of the size of the penalty or the initial penalty on state S_0. We should note from the illustrations of the behaviour of the first and second differences of the Riccati equation how conservative the sufficiency condition of negative definiteness is for both differences. For example, in Figure 6.15b there is no stability at all according to the first difference, stability is achieved from the tenth iteration according to the second difference. And yet the root locus in Figure 6.15a indicates that control remains in the area of stability for all iterations.

Evidently the convergence of the Riccati equation for penalty $Q_u = 1$ is somewhat slower than for $Q_u = 0.001$. This is because the roots of the closed loop are considerably closer to the unit circle.

The next series of illustrations (Figures 6.16 and 6.17) gives similar examples for an unstable system. We should note that the start from $S_0 = 0$ took place in the area of instability and, where higher penalty $Q_u = 1$ is used, more steps are required to attain stabilizing control. As before, the penalty on the finite state $S0 = 1000$ speeded up convergence. In addition to this, the iterations took place in the stable area right from the start.

The last series of illustrations (Figures 6.18–6.21) shows the same examples performed on nonminimum phase system S8. The larger penalty $Q_u = 1$ used in this system produces more satisfactory iteration behaviour, where all the iterations lead to stabilizing control. The root locus corresponding to each iteration was dramatically changed by adding an initial penalty. The behaviour illustrated in Figure 6.21 is especially noteworthy. It can be seen from the root locus that iteration started in the area of stability, soon left it, and then finally returned after step 36. The behaviour of the eigenvalues of the second difference of the Riccati equation is also interesting. In Figure 6.21b it appears to be an indefinite matrix as far as step 36. In this case, the sufficient conditions of stability give the correct region for stability.

6.4.2 The Characteristics of Linear Quadratic Control in the Time Domain

The previous section dealt with stability as the most vital property of control. Users, however, are more interested in the behaviour of closed loop responses. The system output not only depends on input but is also affected by various disturbances. In this section we show in greater depth how to influence disturbance compensation, and how an adaptive controller reacts to a situation where the type of disturbance acting does not correspond to the regression model's definition of disturbance. Later, we will discuss compensation for measurable disturbance where we can make use of a feedforward process. Finally, we will concentrate in more detail on the problems of set point control and observe the transitions to set point changes. The resulting behaviour of the closed loop is then the superposition of cases previously mentioned.

Compensation of Disturbance

There are many types of disturbance which act on the output of the controlled process. It may be a random process or perhaps the response of some filter to a step. If we cannot measure the source of the disturbance, the disturbance must be modelled. In Chapter 3 we showed that the regression model models both system and disturbance, and its characteristics are modelled by filter $\frac{1}{A(z^{-1})}$. Here we consider stochastic disturbance. The characteristics of control behaviour in compensating deterministic disturbance are similar to those of set point control, which is dealt with in the concluding part of this section.

Quadratic criterion synthesis of the controller can only ensure optimal compensation for disturbance modelled by the regression model. It is therefore important that the model represents the characteristics of both system and disturbance. First we look at a situation where the disturbance is characterized by the regression model. We described typical behaviour in Section 6.2, together with the influence of basic parameter Q_u.

In this part we concentrate on two issues:

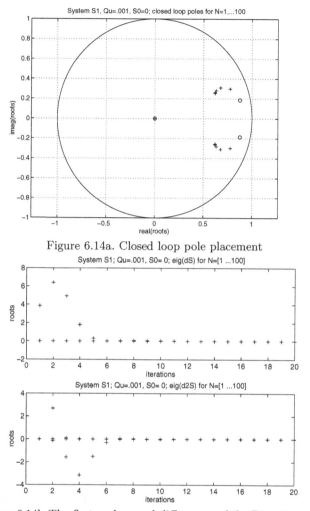

Figure 6.14a. Closed loop pole placement

Figure 6.14b The first and second differences of the Riccati matrix

Figure 6.14. System S1: $Q_u = 0.001, \mathbf{S}_0 = 0$, horizon $T = [1, ..., 100]$

- how does the sampling period affect the quality of compensation?
- what will the control behaviour be when the system is affected by disturbance which differs from that modelled by the regression model?

In our evaluation of the effect of the sampling period we use a continuous-time system with transfer function $G(s) = \frac{B(s)}{A(s)}$ and random disturbance obtained as white noise passes through a filter with transfer function $F(s) = 1/A(s)$. If the noise dispersion is $\sigma_{e_s}^2$, the mean value for the filter output will also be zero and dispersion will be given by

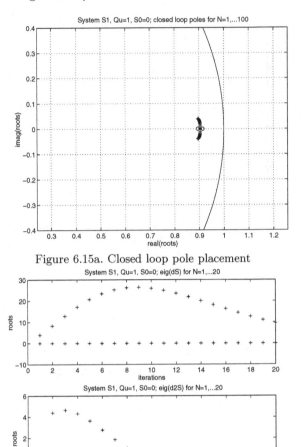

Figure 6.15a. Closed loop pole placement

Figure 6.15b The first and second differences of the Riccati matrix

Figure 6.15. System S1: $Q_u = 1, S_O = 0$, horizon $T = [1, ..., 100]$

$$\sigma_y^2 = \sigma_s^2 \frac{1}{\pi i} \oint \frac{1}{A(s)A(-s)ds} \qquad (6.29)$$

If we consider a discrete controller using a discrete process model in a continuous-time process, then a single continuous-time process is characterized by a set of discrete models, depending on the sampling period chosen. Disturbance dispersion does not change with sampling. However, dispersion of the generating white noise does. This enables one to obtain a relation for the dispersion of actuating noise in the form

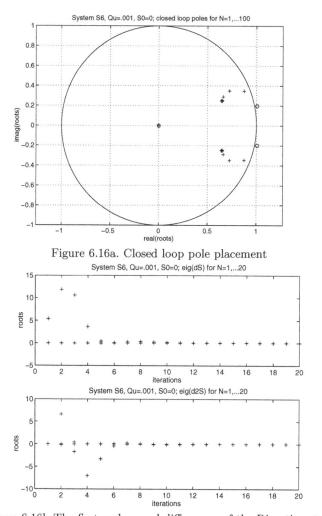

Figure 6.16a. Closed loop pole placement

Figure 6.16b The first and second differences of the Riccati matrix

Figure 6.16. System S6: $Q_u = 0.001$, $S_O = 0$, horizon $T = [1, ..., 100]$

$$\sigma_{e_s,d}^2 = \pi i \frac{\sigma_y^2}{\oint \frac{1}{A(z)A(z^{-1})} \frac{dz}{z}} \qquad (6.30)$$

It can be shown how its size is affected by the sampling period. We take as an example, system (filter) $F(s) = 1/(s+1)^2$. Table 6.1 gives the discrete transfer functions of the filter which correspond to the value of the integral, and resulting dispersion $\sigma_{e_s}^2$ for $T_0 = 0.1\,\text{s}$, $0.2\,\text{s}$, $0.5\,\text{s}$ and $1\,\text{s}$.

Result 1

Dispersion $\sigma_{e_s,d}^2$ of the generating white noise will fall as the frequency of

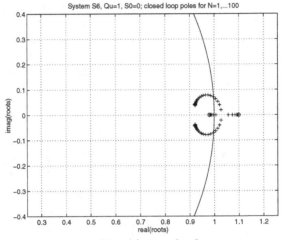

Figure 6.17a. Closed loop pole placement

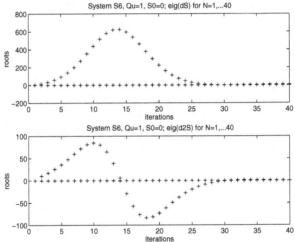

Figure 6.17b The first and second differences of the Riccati matrix

Figure 6.17. System S6: $Q_u = 1, S_O = 0$, horizon $T = [1, ..., 100]$

Table 6.1. Values for white noise dispersion in relation to the sampling period

$T_0[s]$	Filter transfer function	Integral value	Noise dispersion
0.1	$\dfrac{1}{z^{-2} -1.81z^{-1}+ 0.819}$	305.3	0.0033
0.2	$\dfrac{1}{z^{-2}- 1.63z^{-1}+ 0.67}$	46.614	0.0215
0.5	$\dfrac{1}{z^{-2}- 1.21z^{-1}+ 0.368}$	5.4156	0.1847
1	$\dfrac{1}{z^{-2}- 0.736z^{-1}+ 0.135}$	1.7562	0.5694

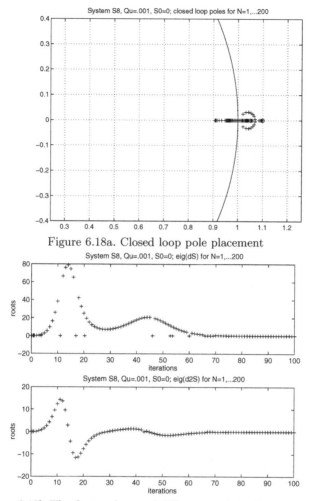

Figure 6.18a. Closed loop pole placement

Figure 6.18b The first and second differences of the Riccati matrix

Figure 6.18. System S8: $Q_u = 0.001, S_O = 0$, horizon $T = [1, ..., 100]$

sampling increases (T_0 gets shorter) if continuous-time random disturbance is represented by a discrete model for various sampling periods.

Result 2

Since $\sigma_{e_s d}^2$ is the lower boundary of attainable values for the control criteria, better disturbance compensation can be achieved by increasing the sampling period.

These results will be tested in simulation according to the diagram given in Figure 6.22.

The two continuous-time systems seen here are modelled by the usual method involving integrators and feedback. The aim of *Transfer function 1* is

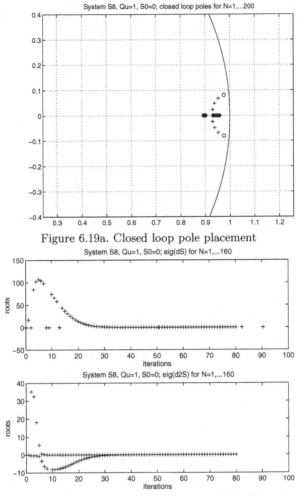

Figure 6.19a. Closed loop pole placement

Figure 6.19b The first and second differences of the Riccati matrix

Figure 6.19. System S8: $Q_u = 1, S_O = 0$, horizon $T = [1, ..., 100]$

to generate disturbance. Suitable random disturbance is obtained by feeding a random signal into the system input.

We have used a *limited spectrum random signal* which is, in effect, a realization of discrete white noise. We can attain the kind of behaviour which imitates reality by manipulating the bandwidth of random noise, the scale, and the length of the graph records. This signal is added to the system output, where it simultaneously represents uncontrolled output. *Transfer function* represents the controlled process itself, which is, again, continuous-time – as it is in reality.

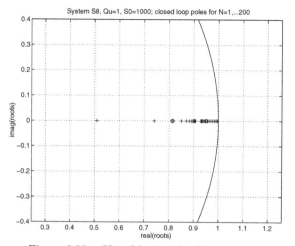

Figure 6.20a. Closed loop pole placement

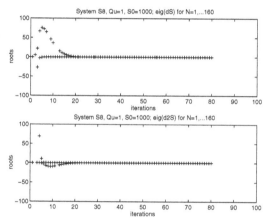

Figure 6.20b The first and second differences of the Riccati matrix

Figure 6.20. System S8: $Q_u = 1, S_O = 1000$, horizon $T = [1, ..., 100]$

Since we are using an LQ controller based on a regression model of the process, the filter and system must be set so as to meet the conditions for representing the process as a regression model. This is why both *Transfer function* and *Transfer function1* have the same denominator.

A further condition, which is that the filter numerator should equal 1, cannot be met in all sampling periods. In fact, it is difficult to derive its discrete transfer function because the input and output are generally sampled at different periods. It helps if we obtain the discrete (regression) model directly from the identification data, rather than transforming from the Laplace to the Z-image. This is how it would normally be done in practice.

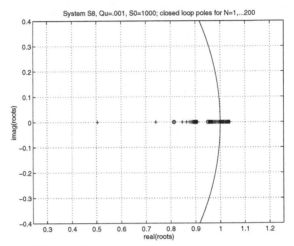

Figure 6.21a. Closed loop pole placement

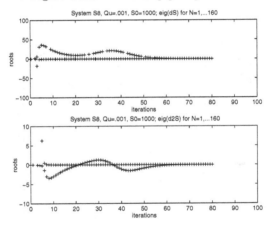

Figure 6.21b The first and second differences of the Riccati matrix

Figure 6.21. System S8: $Q_u = 0.001$, $S_O = 1000$, horizon $T = [1, ..., 100]$

The discrete blocks representing the controller (*i.e.* identification), LQ synthesis, and the controller itself, will all have an optional sampling period. Each image also illustrates the dispersion of model's prediction error and this represents the minimum possible value of the dispersion of disturbance compensation.

Conclusions

1. It is clear from the behaviour seen in Figures 6.24–6.26 that the quality of disturbance compensation depends on the sampling period. A small penalty on u, $Q_u = 0.001$ was used throughout the examples. The relationship between the quality of compensation and the sampling period is especially obvious

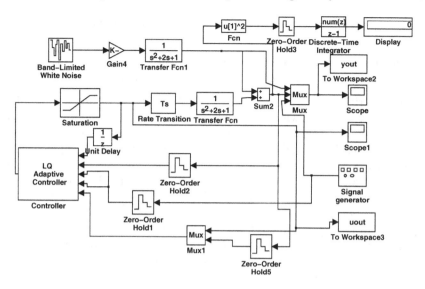

Figure 6.22. Diagram of the simulation of noise compensation

when a small value is chosen.

2. The size of the controller output is independent of the sampling period. More frequent sampling gives better compensation without increasing the amplitude of the input. Naturally, however, the input frequency spectrum is higher.

Let us turn now to the problem of compensation of disturbance where the properties are not characterized by the regression model. Two possibilities must be taken into account:

- a fixed LQ controller
- an adaptive controller.

The case of a fixed LQ controller is relatively easy to deal with. Simply, it will not provide optimal compensation of a disturbance. The method of compensation is governed by the properties of the closed loop created by the system and controller. The stability is not affected.

Adaptive LQ control case is more complicated. The mismatch between the characteristics of the disturbance and those represented by a filter forces the identification process to find coefficients $\bar{a}_i$ of $\frac{1}{A(z-1)}$ so that it predicts the disturbance as efficiently as possible. This, of course, also modifies the transfer function of the system model. The behaviour of a controller based on such parameters cannot be determined; it can only be estimated through an analysis of the robustness of the controller.

We can now demonstrate this kind of situation on some simple examples. Disturbance will be generated by a filter which has a transfer function other than $\frac{1}{A(z)}$.

We have chosen two extreme cases for system S1 to demonstrate its behaviour.

1. Factor $1 - z^{-1}$ has been added to the dynamic of the noise filter.
2. Factor $1 - 0.98z^{-1}$ has been added to the filter numerator.

In the series illustrated by Figure 6.26 we can see an example of standard noise compensation, in which the conditions for the regression model are satisfied. This is followed by three examples where noise is generated according to point 1. Clearly this noise is "slower". In Figure 6.27b it is compensated by an adaptive controller using a second-order model. The estimated parameters differ significantly from those of the system itself. Figure 6.27c features compensation performed by a third-order controller, where the structure of the model permits the inclusion of the entire dynamic of the filter. Compensation is perfect. In the final figure, a fixed controller designed using the system parameters is used for compensation. Compensation here is rather worse than for the adaptive controller using the same structure. These results are given in numeric form in Table 6.2.

Table 6.2. Results of simulation testing

Figure	Identified parameters $\frac{B}{A}$	Loss
6.26b	$\dfrac{0.0047z^{-2}\ 0.0044z^{-1}}{z^{-2} - 1.816z^{-1} + 0.804z^{-2}}$	0.049
6.26c	$\dfrac{0.0047z^{-3} - 0.0003z^{-2} - 0.0044z^{-1}}{z^{-3} - 2.815z^{-2} + 2.6373z^{-1} - 0.82}$	0.0366
6.26d	fixed parameters	0.0519

The second situation is documented in the series of illustrations in Figures 6.27. Figures 6.27a and 6.27c show disturbance and system output. Disturbance here is clearly "fast" and the system is unable to compensate it adequately. The failure of the regression model to meet the assumption of noise results in estimated parameters which differ from those of the system. A controller designed using these parameters is still better than one designed using the system parameters. The character of compensation is roughly the same, but the controller which uses estimated parameters has a noticeably smaller controller output (Figure 6.28b) compared to the one which was designed using the system parameters (Figure 6.28d). The simulation results are given in Table 6.3.

Shifting the parameters alters the system model, with the potential danger that instability will arise due to the difference between the model used in the design and the true system. We will discuss this point in the section dealing with robustness. So as to avoid potential loss of stability we must:

- use a model structure which permits more complex disturbances to be modelled (a higher-order model);

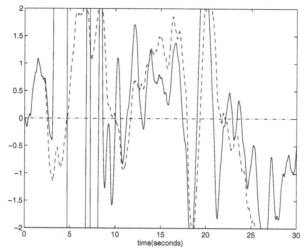

Figure 6.23a. Output (solid line) and disturbance (dotted line)

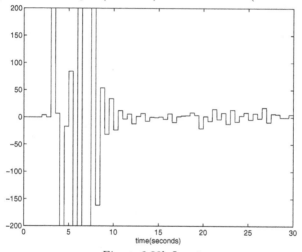

Figure 6.23b Input

Figure 6.23. System S1:$T_0 = 0.5s$, Euler, $\hat{\sigma}_{e_s} = 0.6454$

Table 6.3. Results of simulation testing

Figure	Identified parameters $\frac{B}{A}$	Loss	$\sum u^2$
6.27a	$\frac{0.0055z^{-2} + 0.0069z^{-1} + 0.0034}{z^{-2} - 1.026z^{-1} 0.145}$	1.67	3.83e3
6.27c	fixed parameters	1.58	35 900

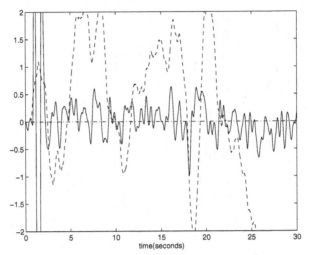

Figure 6.24a. Output (solid line) and disturbance (dotted line)

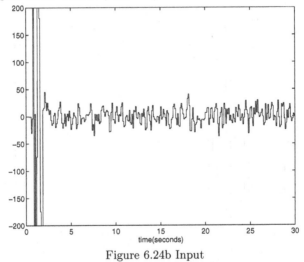

Figure 6.24b Input

Figure 6.24. System S1: $T_0 = 0.1$ s, Euler, $\hat{\sigma}_{e_s} = 0.0983$

- filter the data. This, of course, assumes a knowledge of the character of the disturbance, in order to determine the filter.

A combination of both approaches can be employed.

Figure 6.28 illustrates a SIMULINK® diagram in which the signals are filtered for identification. The use of the filters ensures that the identified parameters represent only some of the characteristics of the process. For example: an integration factor is often added to the open-loop to ensure zero steady state error. This factor will appear in the estimated parameters. This

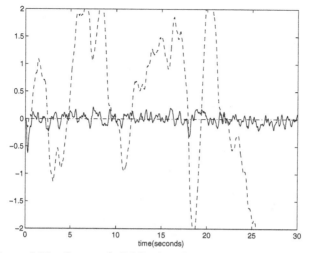

Figure 6.25a. Output (solid line) and disturbance (dotted line)

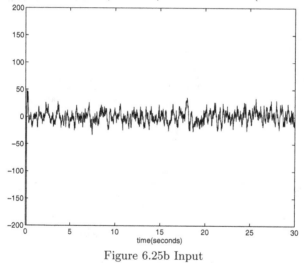

Figure 6.25b Input

Figure 6.25. System S1: $T_0 = 0.01$ s, Euler, $\hat{\sigma}_{e_s} = 0.0185$

means the disturbance filter will contain the factor as well. This dramatically changes the noise properties. For a better solution see Section 6.5.1.

The External Measurable Variable

If we can measure the source of disturbance $v(k)$, which generates the disturbance at the output (Figure 3.1) ,

$$y_v = \frac{D(z^{-1})}{A(z^{-1})}$$

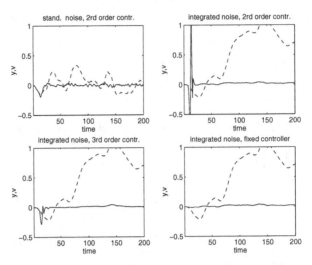

Figure 6.26. Compensation of a disturbance where the filter denominator is expanded by $1 - z^{-1}$

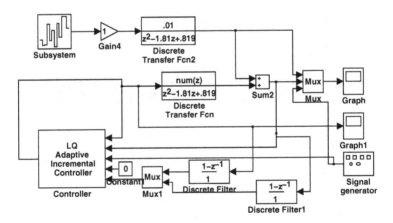

Figure 6.27. Compensation of disturbance where the filter numerator has been expanded by $1 - 0.98z^{-1}$

it is a good idea to use this knowledge in compensation. In practice, disturbance can commonly be measured, since it need not be true disturbance. We can use any signal connected in some way with the system output if its influence can be described by filter $\frac{D(z^{-1})}{A(z^{-1})}$. The source of the disturbance in this case need not be white noise but any signal generated by filter

$$F_v = \frac{1}{A_v(z^{-1})}$$

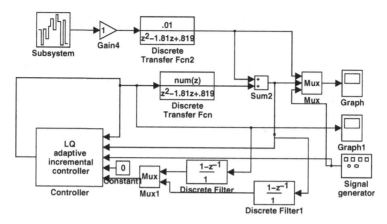

Figure 6.28. SIMULINK® diagram using filters for identification

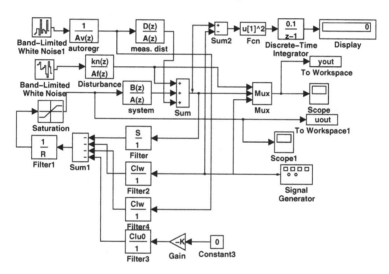

Figure 6.29. SIMULINK® diagram of the compensation for measurable disturbance

Coefficients av_i of this filter now represent the dynamic of the measurable disturbance. We either know these, or can identify them in the same way as we identify the system parameters.

The SIMULINK® diagram in Figure 6.29 will be used to demonstrate the compensation of measurable disturbance. In this diagram the measurable disturbance is generated randomly by filter F_v, shown above. The value is measured and used in the next feedforward component of the controller with transfer function $\frac{Cld(z^{-1})}{R(z^{-1})}$. The output disturbance is governed by the regression model as it passes through filter $\frac{D(z^{-1})}{A(z^{-1})}$.

Nonmeasurable disturbance is suppressed in the first three responses shown in Figure 6.30 so as to highlight the compensation of measurable disturbance. Figure 6.30a shows a process in which disturbance is not being measured; Figure 6.30b one where the disturbance is measured but we are unfamiliar with its properties (with the model which generates it); Figure 6.30c shows the situation where the disturbance is fully understood, and Figure 6.30d shows the overall behaviour with measured and unmeasured disturbance and set point transition.

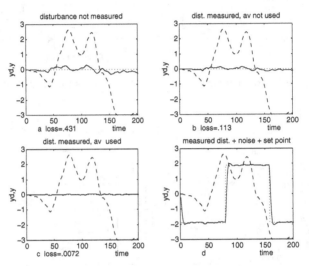

Figure 6.30. Different forms of compensation for external disturbance, (- -) uncompensated disturbance

Set point Tracking

Set point tracking is one of the most important functions of the control loop. The quality of control is often judged on the form of response to the typical changes of the set point (steps – the step response). When we are designing a controller to track the set point we find the following differences with disturbance compensation:

- The transfer function from disturbance to output differs from the transfer function from set point to output.
- A further signal must be taken into account in the minimization criterion and, in addition, we must know its future values.

Unlike most of the other signals in the loop, a knowledge of the future behaviour of the set point is, in practice, very common and natural. Moreover it can almost always be applied simply by delaying the true change in the set point over several (tens of) sampling periods.

The standard approach to optimization, in which the unknown values of the signals in the criterion are predicted using a model, can always be used. This approach can also be applied to set point control. The model used to predict the future constant set point is

$$w(k) = w(k-1) \tag{6.31}$$

Let us discuss in greater depth the situation where the future desired output values are known. If we look at the diagram of the controller (Figure 6.2), we can see that optimization generates both the feedback and feedforward parts of the controller. It is important to remember that, in an adaptive controller, the feedforward component is an integral part of the control transfer function, and any discrepancy between model and reality may be reflected in an error in set point tracking. The filter transfer function in the feedforward component is

$$\frac{u}{w} = \frac{F(z)}{R(z^{-1})} \tag{6.32}$$

when $F(z) = f_0 + f_1 z + f_2 z^2 + \ldots + f_{nw} z^{nw}$

Transfer function (6.32) is noncausal, as expected. The coefficients of the transfer function are obtained using the optimization process (see Section 6.7).

If the control signal is constant, the transfer function can be simplified togive

$$\frac{u}{w} = \frac{\sum_{i=1}^{nw} f_i}{R(z^{-1})}.$$

Term $\sum_{i=0}^{nw} f_i$ is obtained using the optimization process, by accumulating all future set points into one column of matrix $\mathbf{S}$. It is simpler to use filter (6.32) immediately if other types of future behaviour are known, even though it is possible to obtain specific results for each given signal. We will try to demonstrate this using a ramp example. When the ramp has initial value w_0 and increment δw the filter output can be written as

$$u_w = \frac{\sum_{i=1}^{nw} f_i}{R(z^{-1})} w_0 + \frac{\sum_{i=1}^{nw} i f_i}{R(z^{-1})} \delta w$$

In the optimization procedure this is done by giving the Riccati matrix an extra column which corresponds to the increment on the ramp.

Figure 6.31 shows the simulation plan for experiments on set point transitions, including pre-programming. Since SIMULINK® is unable to generate a vector of future values, it must work with delayed real values and compare the system output with the delayed set point. The delay is effected in the *oldval* block. The product $\sum_{i=1}^{nw} f_i w_{t+i}$ is performed in the block *trail1*. The remaining blocks are standard. The controller parameters with pre-programming are calculated by the *lqex3.m* procedure which consists of calculating steady optimal control, and subsequently the pre-programming component. Otherwise the form of the pre-programming filter would be affected by the initial conditions for the Riccati equation. Adaptive controllers with a pre-programming property are contained in the SIMULINK® scheme *adlqsw* or those containing "w" in its name. Let us now look at the typical behaviour of a simple S1 system.

Figure 6.32 shows the step response. The corresponding reaction of the controller output can be seen in Figure 6.33. The coefficients of the pre-programming filter are plotted in Figure 6.34, where we see that $f_i = 0$ for $i > 16$. Therefore, if the pre-programming length is made 10 (see Figure 6.35), the steady state value will be greater than the set point because we neglected the negative coefficients for $i > 10$. Similarly, when the pre-programming length is 5 (Figure 6.36), the output does not reach the required value. In the remaining figures 6.37–6.39 we illustrate output, input and coefficients f_i for a controller using a different penalty on the input. Figures 6.40–6.42) show control of nonminimum phase systems.

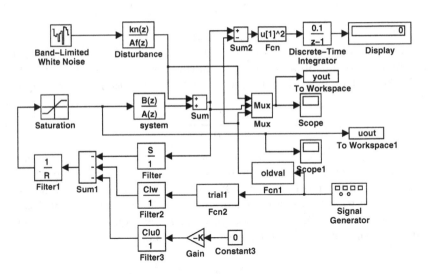

Figure 6.31. Simulation with pre-programming

Conclusions
Pre-programming uses information on future control and so achieves optimal response both from the point of view of output quality and demands placed on the input.
Zero Steady State Error
A typical drawback of LQ design using minimization criterion (6.2) is the nonzero steady state error in the step response when the system contains no integrator term. This offset is particularly obvious when, for whatever reason,

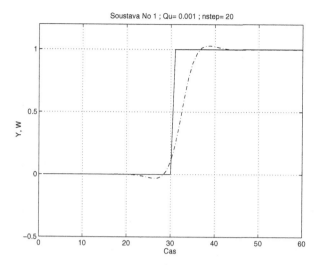

Figure 6.32. S1 output, 20 steps ahead strategy

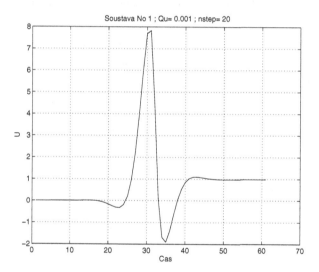

Figure 6.33. S1 input, 20 steps ahead strategy

a higher penalty Q_u must be used. We have already met with this feature in Section 6.2

The standard solution to this problem is to add an integrator to the open-loop. However this changes the transfer function of the disturbance filter as well and, by doing so, also changes its assumed character. Though the response to the transition improves, the disturbance compensation may deteriorate. We must use an ARMAX model and assume that $C = z - 1$ so that the addition of an integrator does not change the compensation of disturbance.

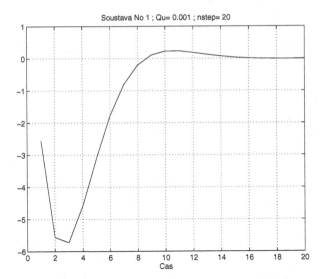

Figure 6.34. S1: coefficients f_i 20 steps ahead strategy

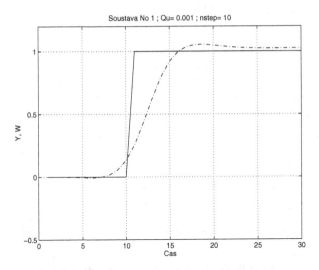

Figure 6.35. S1: output, 10 steps ahead strategy

The steady state error (offset) can be eliminated by penalizing the increment on u, *i.e.* penalizing $\triangle u$ itself. This penalty does not limit the size of the controller output, only changes in it.

Another solution can be derived from criterion (6.3) where we take account of another signal: the reference input. In reality it is more natural not to penalize the entire controller output, only the part remaining after subtraction

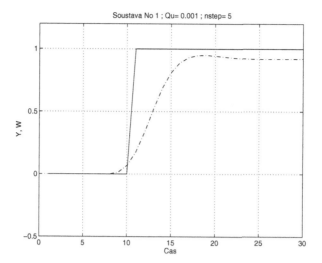

Figure 6.36. S1: output, 5 steps ahead strategy

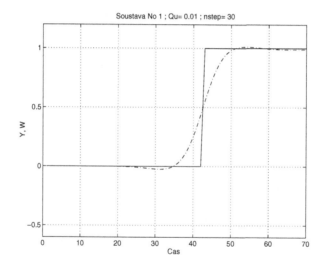

Figure 6.37. S1: output, 30 steps ahead strategy

of the value of u needed to achieve the required output level. The required size of $u0$ can be obtained in two ways:

- $u0$ can be taken as another variable to be used in the minimization criterion (see [124]).
- $u0$ is a signal which is proportional to the set point and must be added to $u(k)$ to compensate for the offset.

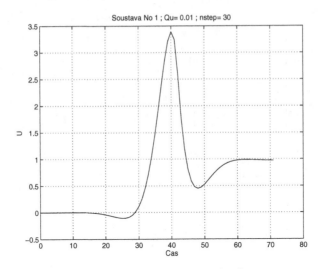

Figure 6.38. S1: input, 30 steps ahead strategy

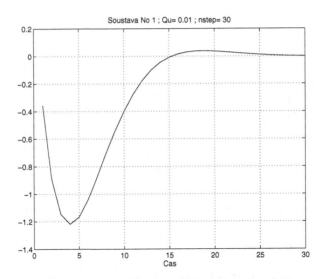

Figure 6.39. S1: coefficients f_i 30 steps ahead strategy

The second method is easier to interpret and is used, *e.g.* in SIMULINK®
diagram *schema1p.m*, illustrated in Figure 6.43. Signal $u0$ will be used for
other purposes in the following section.

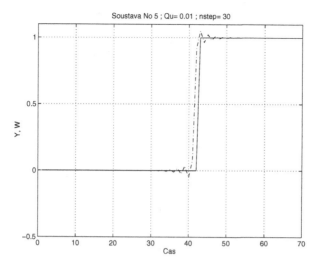

Figure 6.40. Nonminimum phase system S6: output

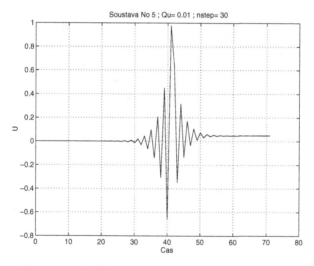

Figure 6.41. Nonminimum phase system S6: input

6.4.3 The Characteristics of Linear Quadratic Control in the Frequency Domain

We know that in a linear system the properties of a control loop with regards to time (transfer function characteristic) and frequency are two sides of the same coin. Until now we have concentrated on the time aspect. The regression model describes the system over time, the minimization criterion was performed in the time domain, and we have demonstrated the typical characteristics of compensation of disturbance and set point tracking. These

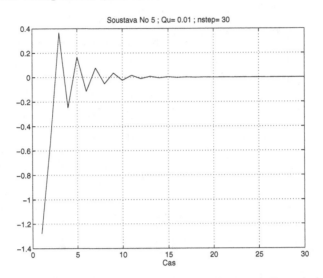

Figure 6.42. Nonminimum phase system S6: coefficients f_i

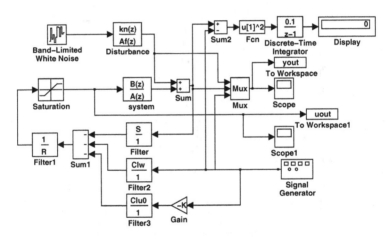

Figure 6.43. Compensation of an offset by signal $u0$

are of primary importance in adaptive control. We will now deal with the other side of the question, which is LQ control with regard to frequency. It is particularly important to observe these properties so as to be able to evaluate the stability of the loop and the robustness of the design. This involves an analysis of the behaviour of a system which is different from the model used in the control design. With reference to frequency, it will also be simpler to demonstrate the effect of the sampling period on stability and the quality of control behaviour. When illustrating the frequency characteristic of discrete systems obtained from sampling a continuous-time system, attention must be paid to the correct transformation of frequency. The discrete transfer function only represents the continuous frequency over half the sampling frequency. This means that sampling period $T_0 = \frac{1}{f_0}$ can reflect the maximum angular frequency

$$\omega_m = 2\pi \frac{f_0}{2} = \pi f_0 = \frac{\pi}{T_0} \tag{6.33}$$

This frequency is transformed by sampling to discrete angular frequency $\omega_{d,m} = \pi$. Therefore, if we are dealing with a control loop in which the discrete transfer function was obtained through sampling, we cannot avoid the relation between the discrete and continuous frequencies arising out of (6.33)

$$\omega_d = \omega_s T_0$$

Sometimes the discrete frequency is expressed as a ratio to the Nyquist frequency, which is half the sampling frequency $f_r = \frac{\omega_d}{\omega_{d,m}}$. In this case the highest frequency has a value of 1. The continuous frequency is obtained using $f_s = f_r f_0 / 2$, and the discrete angular frequency using $\omega_d = f_r \pi$.

The frequency characteristics of LQ controller loops display several typical properties:

- the frequency characteristic of an LQ controller has a tendency to amplify signals at higher frequencies,
- the frequency characteristic of an open-loop in the complex plane (the Nyquist diagram) displays typical behaviour around points (-1, 0) of the complex plane,
- the LQ controller attempts to maintain the frequency characteristic of a closed loop transfer function between output and set point in the logarithmic coordinates flat to the highest frequency.

The Frequency Characteristics in Logarithmic Coordinate

The frequency characteristics in logarithmic coordinates are useful for demonstrating the effects of the sampling period and penalization. We can show the typical behaviour of a controller on an example of control of a continuous-time system with transfer function

$$G(p) = \frac{1}{(p+1)^3}.$$

When the sampling periods are $T_0 = 0.1, 0.2, 0.5$ and 1s, the discrete transfer functions are marked as S10, S12, S15 and S20 in the *model.m* file. Figure 6.44 illustrates the frequency characteristics of those continuous-time and discrete transfer functions corresponding to each individual sampling period. We have purposely emphasized the maximum frequency where the discrete model still represents the continuous-time system at different sampling periods. We achieve a match with the continuous frequency characteristic up to

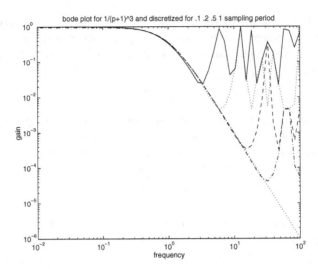

Figure 6.44. The frequency characteristic of a discretely modelled continuous-time system at various sampling periods

frequency $\frac{\pi}{T_0}$; at higher frequencies the value of the characteristic periodically repeats itself. Due to the logarithmic coordinates used and the paucity of points at high frequency, the periodicity, and particularly symmetry, of the solution are distorted.

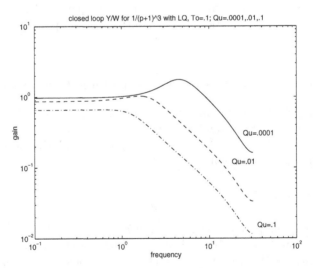

Figure 6.45. The frequency characteristic of a closed loop with various penalties

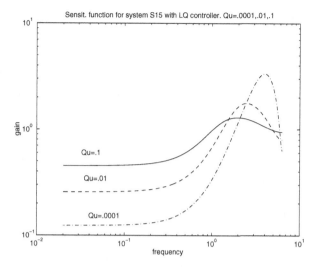

Figure 6.46. The sensitivity function for various penalties

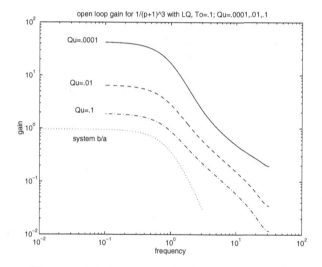

Figure 6.47. Open-loop gain for various penalties

Figure 6.45 shows the system's set point closed loop transfer function frequency characteristic for different penalties and sampling period $T_0 = 0.1$ s.

Figure 6.46 illustrates the sensitivity function (disturbance transfer function) of the same system, with sampling period $T_0 = 0.1$ s. Penalty Q_u shifts the controller frequency characteristic vertically and so significantly alters the total gain of the open-loop (see Figure 6.47).

In other types of system, for example nonminimum phase, the shift in the controller frequency characteristic is markedly smaller.

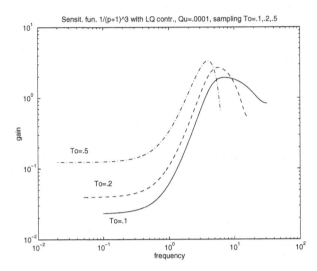

Figure 6.48. The sensitivity function for various sampling periods

Figure 6.48 illustrates the changing character of the sensitivity function where the system uses an LQ controller obtained byoptimization of the same criterion, but different sampling periods.

The Frequency Characteristic in the Complex Plane

The frequency characteristics of an open-loop transfer function with an LQ controller show typical behaviour by surrounding the point $(-1, 0)$ in the complex plane.

The form of the frequency characteristic can be deduced from the frequency interpretation of the Riccati equation and, as a matter of interest, we give this procedure in detail in [121].

Our starting point is the standard form of the algebraic (steady state) Riccati Equation (6.8). Frequency interpretations almost always apply to the steady state. Omitting index i, we use relation (6.9) to obtain

$$\mathbf{S} = \mathbf{F}^T \mathbf{S} \mathbf{F} + \mathbf{Q}_x - \mathbf{L}^T(\mathbf{Q}_u + \mathbf{G}^T \mathbf{S} \mathbf{G})L \qquad (6.34)$$

We then move the term $\mathbf{F}^T \mathbf{S} \mathbf{F}$ to the left side of the equation and simultaneously add and subtract terms $z\mathbf{I}\mathbf{F}^T\mathbf{S}$ and $z^{-1}\mathbf{I}\mathbf{F}\mathbf{S}$ on this side. The left side can then be manipulated as follows:

$$(z^{-1}\mathbf{I} - \mathbf{F}^T)\mathbf{S}(z\mathbf{I} - \mathbf{F}) + (z^{-1}\mathbf{I} - \mathbf{F}^T)\mathbf{S}\mathbf{F} + \mathbf{F}^T\mathbf{S}(z\mathbf{I} - \mathbf{F})$$

The equation is then multiplied from the left by $(z^{-1}\mathbf{I} - \mathbf{F}^T)^{-1}$, and from the right by $(z\mathbf{I} - \mathbf{F})^{-1}$, to obtain

$$\mathbf{S} + \mathbf{S}\mathbf{F}(z\mathbf{I} - \mathbf{F})^{-1} + (z^{-1}\mathbf{I} - \mathbf{F}^T)^{-1}\mathbf{F}^T\mathbf{S} =$$

$$(z^{-1}\mathbf{I} - \mathbf{F}^T)^{-1}\mathbf{Q}_x(z\mathbf{I} - \mathbf{F})^{-1} - z^{-1}\mathbf{I} - \mathbf{F}^T)^{-1}\mathbf{L}^T(\mathbf{Q}_u + \mathbf{G}^T\mathbf{SG})\mathbf{L}(z\mathbf{I} - \mathbf{F})^{-1}$$

The final term of the right-hand side is moved to the left, both sides are multiplied by $\mathbf{G}^T$ from the left and G from the right, and $\mathbf{Q}_u$ is added to both sides. We use relation $(\mathbf{Q}_u + \mathbf{G}^T\mathbf{SG})\mathbf{L} = \mathbf{G}^T\mathbf{SF}$ to obtain successively

$$\mathbf{Q}_u + \mathbf{G}^T(z^{-1}\mathbf{I} - \mathbf{F}^T)^{-1}\mathbf{Q}_y(z\mathbf{I} - \mathbf{F})^{-1}\mathbf{G} =$$
$$\mathbf{Q}_u + \mathbf{G}^T\mathbf{SG} - (\mathbf{Q}_u + \mathbf{G}^T\mathbf{SG})\mathbf{L}(z\mathbf{I} - \mathbf{F})^{-1} - (z^{-1}\mathbf{I} - \mathbf{F}^T)^{-1}\mathbf{L}^T$$
$$(\mathbf{Q}_u + \mathbf{G}^T\mathbf{SG}) + \mathbf{G}^T(z^{-1}\mathbf{I} - \mathbf{F}^T)^{-1}\mathbf{L}^T(\mathbf{Q}_u + \mathbf{G}^T\mathbf{SG})\mathbf{L}(z\mathbf{I} - \mathbf{F})^{-1}\mathbf{G}$$
$$= (\mathbf{Q}_u + \mathbf{G}^T\mathbf{SG}) - (\mathbf{Q}_u + \mathbf{G}^T\mathbf{SG})\mathbf{L}(z\mathbf{I} - \mathbf{F})^{-1} - (z^{-1}\mathbf{I} - \mathbf{F}^T)^{-1}\mathbf{L}^T$$
$$(\mathbf{Q}_u + \mathbf{G}^T\mathbf{SG}) + \mathbf{G}^T(z^{-1}\mathbf{I} - \mathbf{F}^T)^{-1}\mathbf{L}^T(\mathbf{Q}_u + \mathbf{G}^T\mathbf{SG})\mathbf{L}(z\mathbf{I} - \mathbf{F})^{-1}\mathbf{G}$$
$$= [\mathbf{I} - \mathbf{L}(z^{-1}\mathbf{I} - \mathbf{F}^T)^{-1}\mathbf{G}]^T(\mathbf{Q}_u + \mathbf{G}^T\mathbf{SG})[\mathbf{I} - \mathbf{L}(z^{-1}\mathbf{I} - \mathbf{F}^T)^{-1}\mathbf{G}] \quad (6.35)$$

For a single-input single-output case this can be written as

$$\frac{Q_u}{\mathbf{G}^T\mathbf{SG} + Q_u} + \frac{|(z\mathbf{I} - \mathbf{F})^{-1}\mathbf{G}|^2}{\mathbf{G}^T\mathbf{SG} + Q_u} = |\mathbf{I} - \mathbf{L}(z^{-1}\mathbf{I} - \mathbf{F}^T)^{-1}\mathbf{G}|^2 \quad (6.36)$$

Therefore

$$\frac{Q_u}{\mathbf{G}^T\mathbf{SG} + Q_u} < |\mathbf{I} - \mathbf{L}(z^{-1}\mathbf{I} - \mathbf{F}^T)^{-1}\mathbf{G}|^2 \quad (6.37)$$

Inequality (6.37) can be interpreted simply: the distance of the frequency characteristic from point $(-1, 0)$ in the complex plane is always greater than the constant which appears on the left side of the inequality.

These relations required relatively complicated manipulation of the Riccati equation. Similar relations which also include the case of a dynamic output controller can be obtained far more easily from the system transfer function and the polynomial synthesis of an LQ controller [106]. We will demonstrate this on a simplified example in which synthesis is performed in two stages:

1. First the polynomial of the closed loop ϕ is calculate from the factorization equation

$$Q_y\bar{B}B + Q_u\bar{A}A = \phi_0^2\bar{\phi}\phi \quad (6.38)$$

2. The polynomial equation is solved

$$AR + BS = \phi$$

$\bar{B}, \bar{A}$ in the equation denote the conjugated polynomial for numerator B and denominator A of the system transfer function, and ϕ_0 is the normalization coefficient which yields $\phi = 1 + p_1 z^{-1} + \ldots + p_n z^{-n}$. If we divide (6.38) by terms $\bar{A}A, \bar{R}R$ and ϕ_0^2, where R is the denominator of the controller transfer function, we obtain equation

$$\frac{Q_u}{\phi_0^2\bar{R}R} + Q_y\frac{\bar{B}B}{\phi_0^2\bar{A}A\bar{R}R} = \frac{\bar{\phi}\phi}{\bar{A}A\bar{R}R} \quad (6.39)$$

or

$$\frac{Q_u}{\phi_0^2 \bar{R}R} + Q_y \frac{\bar{B}B}{\phi_0^2 \bar{A}A\bar{R}R} = |1 + G|^2$$

The inverse of the module of the sensitivity function is on the right side. On the left side we are principally interested in the first expression. It follows from the algorithm of the solution to the Riccati equation (Section 6.7) that

$$\phi_0^2 = Q_u + \mathbf{P}_u^T \mathbf{S} \mathbf{P}_u$$

and that this value equals the $\mathbf{H}_{uu}$ element of minimized matrix $\mathbf{Hn}$ (see Section 6.7). Unlike the state space example (6.36), where the minimum distance of the frequency characteristic from point $(-1, 0)$ in the complex plane is limited by the first term on the left side of Equation (6.30), in Equation (6.39) the minimum distance is also a function of $\frac{1}{\bar{R}R}$. We cannot therefore claim a guaranteed minimum distance.

Robustness in LQ Controllers

Robustness has gained popularity in recent years but, more importantly, it is a vital attribute of all controllers for practical applications. The reason for this is simple. The perfect match between model and true system, assumed in controller design, cannot be guaranteed in practice. We do have methods to design a so-called robust controller, but these usually result in a controller set to maintain a certain quality of control for a whole class of systems which differ from the model in a defined way. Quality, however, is often only average. Here we overcome the problems of insufficient knowledge of the real system using an adaptive approach. Notwithstanding, the properties which determine robustness must be taken into account because adaptation relying on the identification of model parameters can never ensure a perfect match. We use results taken from our analysis of frequency response to observe robustness.

When treating robustness properties, we will mainly consider the stability of the system closed loop for which the controller was not originally designed.

The size of changes in the system which can occur without destabilizing the loop is easily determined from the shortest distance between the open-loop frequency characteristic and point $(-1, 0)$. This is illustrated in Figure 6.49, which shows

the open-loop frequency characteristic. Point P of the frequency characteristic is the closest to point K $(-1, 0)$. The following vector relation applies in the complex plane

$$\mathbf{KP} = \mathbf{K0} + \mathbf{0P} \tag{6.40}$$

or

$$P(j\omega) = I + G(j\omega) = \frac{AR + BS}{AR}$$

Since the final expression is the inverse of the sensitivity function, the inverse of $|P(j\omega)|$ yields the modulus of the sensitivity function. In the previous section we showed how the minimum distance between the open-loop frequency

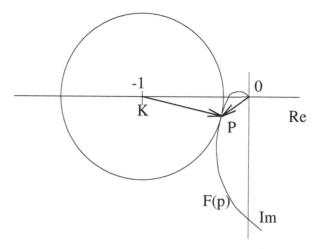

Figure 6.49. Evaluating robustness from the frequency characteristic

characteristic and critical point $(-1, 0)$ on the complex plane can be found through optimization. Therefore the result of optimization can guarantee a certain level of robustness.

Note:

In other words: the frequency characteristic does not intersect with the circle centred on point (-1, 0) of the complex plane and radius $\frac{Q_u}{(G^T SG + Q_u)^{1/2}|R_m|}$, where $|R_m|$ is the maximum value of the frequency characteristic modulus of the controller transfer function denominator.

Unlike similar results for continuous-time systems, the term omitted from Equation (6.36) can have a very positive effect on the radius of the circle since, depending on the chosen sampling period, the modulus of the frequency characteristic may be sufficiently large at point $\omega = \pi$.

The tolerance of a steady state LQ controller to a change in gain in the range $(1/2 - \infty)$ and to a change in phase of 60° is well known. This, however, **does not apply** to discrete systems where the state controller is derived from (6.36). In our example, which is characterized by feedback from the system output and not from the state, the robust stability is characterized by Equation (6.39). This equation affects robustness using penalty Q_u both directly and via controller denominator R. The form of the frequency characteristic is also influenced by the sampling period. We can demonstrate the typical effect of these variables by analyzing a simple example.

Figure 6.50 illustrates the open-loop frequency characteristic of system $G(s) = \frac{1}{(s+1)^3}$, sampled at period $T_0 = 0.1\,\mathrm{s}$ (system S10), with LQ controller for $Q_u = 0.0001$.

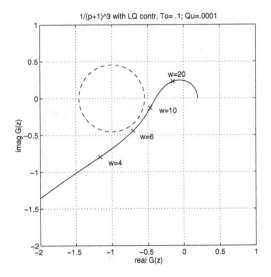

Figure 6.50. The open-loop frequency characteristic, $T_0 = 0.1$ s, $Q_u = 0.0001$

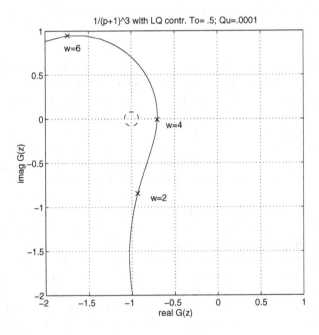

Figure 6.51. The open-loop frequency characteristic , $T_0 = 0.5$ s, $Q_u = 0.0001$

The dotted line marks the circle which the frequency characteristic cannot intersect. Figure 6.51 illustrates a similar example where the sampling period has been changed to 0.5 s.

We will now look at a case where we use an identified model and models of lower order than the actual system in the design of the controller. We again take system $G(s) = \frac{1}{(s+1)^3}$. The identification experiments are used to obtain third-, second- and first-order models, the parameters of which are given in the table below.

Table 6.4. Discrete model parameters

Order	Numerator
	Denominator
Ideal	$0.0001547 + 0.000574z^{-1} + 0.0001331z^{-2}$
	$1. - 2.7145z^{-1} + 2.4562z^{-2} - 0.74080z^{-3}$
3	$0.00037136 + 0.00015823z^{-1} + 0.00006041z^{-2} + 0.00013480z^{-3}$
	$1. - 2.7926z^{-1} + 2.6130z^{-2} - 0.81810z^{-3}$
2	$-0.0000536 + 0.0001094z^{-1} + 0.0006688z^{-2}$
	$1. - 1.987z^{-1} + .99540z^{-2}$
1	$0.000869 + 0.000799z^{-1}$
	$1. - 0.99630z^{-1}$

Figure 6.52 illustrates the amplitude and phase characteristics of these transfer functions. Figure 6.53 shows the open-loop frequency characteristics of a third-order system using a controller designed on the basis of identified first-, second- and third-order models for penalty $Q_u = 0.0001$ s. It is clear that a second-order model can still result in a stable loop, but a first-order model already shows signs of instability. If we increase the penalty in this case the loop will stabilize.

In the section on noise compensation we saw that, where the filter-generating disturbance contained $C(z-1) = 1-0.98z^{-1}$ in the numerator, the parameter estimates differed radically from the true parameters of the system. Looking at this with regards to frequency, it will be shown what effect this has on stability.

Figure 6.54 illustrates the frequency characteristics of a system with idealized and identified parameters. Although the two characteristics differ, those of the open-loop are almost identical, especially at high frequencies. This is illustrated in Figure 6.55.

To a large extent, the stability and quality of control behaviour when the model differs from reality depend upon how close the frequency characteristics

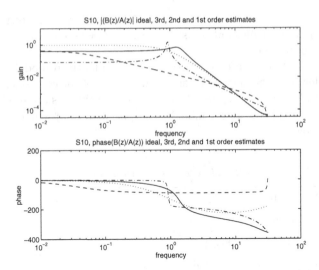

Figure 6.52. The frequency characteristic of an idealized system and third-, second- and first-order models

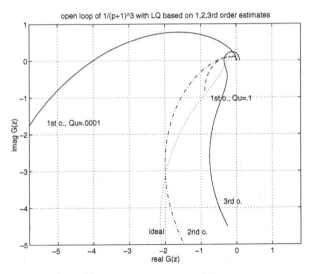

Figure 6.53. The open-loop frequency characteristic using various controllers

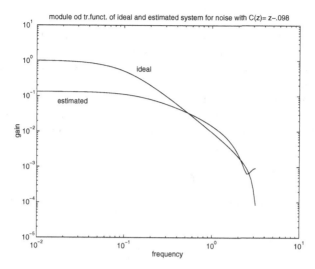

Figure 6.54. The frequency characteristic of an idealized and identified system

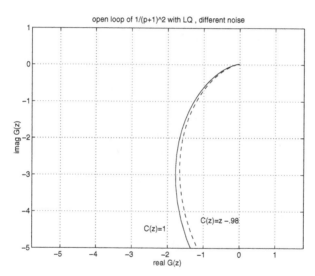

Figure 6.55. The open-loop frequency characteristic in an idealized and identified system

are at higher frequencies. In the first instance we saw that a first-order model corresponds quite closely to the idealized system at lower frequencies but differs merkedly at higher frequencies, resulting in instability. Contrary to this, in the second instance the differences between system and model were greatest at low frequencies.

Monograph [69] gives a relation which may modify the frequency characteristic so that the process remains stable. In the introduction to this section we showed the conditions which apply to possible changes in the open-loop transfer function. These are given by the inequality

$$|G(z) - G0(z)| < |1 - G(z)|$$

where $G(z), G0(z)$ are the open-loop frequency characteristics of a true and idealized system. Since

$$|G(z) - G0(z)| = |\frac{S}{R}||\frac{B}{A} - \frac{B0}{A0}|,$$

is valid we can write

$$|\frac{B}{A} - \frac{B0}{A0}| < |\frac{AR + BS}{AR}\frac{R}{S}|$$

and, after cancelling R, we obtain

$$|\frac{B}{A} - \frac{B0}{A0}| < |\frac{AR + BS}{AS}|$$

If we modify this expression in the way described in [69], where, in our example $H_m = \frac{BClw}{AR+BS}$, $H_{ff} = \frac{Clw}{R}$, $H = \frac{B}{A}$, $H_{fb} = \frac{S}{R}$ we obtain

$$|\frac{B}{A} - \frac{B0}{A0}| < |\frac{H(z)Clw}{H_m(z)S(z)}|.$$

6.5 Tuning an Linear Quadratic Controller

In the previous sections we became acquainted with the relevant properties of a control loop using an LQ controller. This section deals with the issue of adjusting an LQ controller to meet the demands the user makes on control behaviour as fully as possible. In the following passages we will recall useful rules for adjusting and starting up an adaptive controller in a real system.

6.5.1 Tuning a Controller

At first glance it might seem rather curious to tune an LQ optimal controller when the optimization process itself should generate the best control behaviour. Quite simply, although the optimization process really does guarantee optimal control behaviour, provided certain conditions are satisfied, this does not necessarily mean that it is optimal from the user's viewpoint. The behaviour obtained from an optimization criterion with a large penalty will probably not satisfy the user in terms of speed of response and steady state error. The principal on which the tuning of an LQ controller is based consists primarily in choosing the criterion (penalty) which results in the type of control the user wants. We have seen that there are other influences acting on the control loop, particularly the sampling period. The next section is devoted to ways of modifying the criterion.

Adjustable Criterion Parameters

In our previous discussion of LQ controllers we have referred to one of the tuning elements – penalty Q_u, or $Q_{\triangle u}$. It has been shown, however, that, by itself, it is insufficient to ensure full adjustment of behaviour. This can be shown on an example. The root locus of the closed loop roots of system S3, with reference to the size of penalty Q_u, is illustrated in Figure 6.56. It can be seen that, regardless of the penalty, the dominant roots are formed by a complex conjugated pair. Simulation shows that there is always an overshoot in the step response . Penalty Q_u cannot, therefore, be used to achieve the required response without overshoot. Figure 6.57 illustrates the geometric root locus of the roots of the same system, but a further penalty on the output difference $(y(k) - y(k-1))^T Q_{dy}(y(k) - y(k-1))$ was added to the criterion.

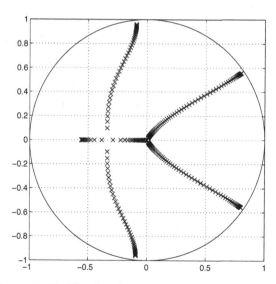

Figure 6.56. The closed loop roots for $0 < Q_u/Q_y < \infty$

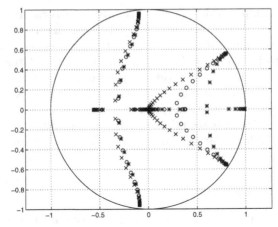

Figure 6.57. The closed loop roots for $0 < Q_u/Q_y < \infty$ and $Q_{dy} = 0; 1; 10$

If penalty Q_{dy} has been well chosen we can achieve a nonoscilatory tran-sient to a closed loop change. This example indicates the necessity of expand-ing the set of adjustable criterion parameters. Here we should note that the classic state space formulation of LQ control permits the closed loop poles to be placed in a sufficiently large region, providing the controllability con-ditions have been satisfied. Superficially it may appear that we have got rid of this possibility when using the input/output formulation of LQ control. When we use input/output penalties, the closed loop roots lie on a specific single parametric curve, the parameters of which are Q_u/Q_y. We know that choosing the roots from this set does not necessarily represent the behaviour required of the loop. However, this is simple to correct both principally and algorithmically.

Generalized Penalization

The pseudo-state solution to the minimization of the quadratic criteria (see Sections 6.1 and 6.7) allows the problem-free expansion of the penalty to include a general penalization on the pseudo- state. Until now we have used penalty matrix

$$\mathbf{Q} = \begin{bmatrix} Q_u & ... & 0 & ... & 0 & 0 \\ 0 & ... & 0 & ... & 0 & 0 \\ 0 & ... & Q_y & ... & 0 & 0 \\ 0 & ... & 0 & & 0 & 0 \\ 0 & ... & 0 & ... & 0 & 0 \\ 0 & ... & 0 & ... & 0 & 0 \end{bmatrix}$$

to penalize input and output; we can use a general $\mathbf{Q}$ matrix in its place.

It is fairly difficult to find a generalized matrix which would result in the required modification to the control behaviour. It is easier to regard such a matrix as the sum of several penalties, each with a simpler matrix. It is a good idea to express each square symmetric matrix as the sum of rank 1 matrices, which are the product of the vector and its transpose. Each component of the penalty then has the form

$$\alpha_i \mathbf{f}_i^T \mathbf{f}_i$$

where α is the weighting of the penalty and $\mathbf{f}_i$ is the numeric vector. Each term of the criterion will have the form

$$\mathbf{Q}i = \mathbf{z}(k)^T \mathbf{f}_i^T \alpha \mathbf{f}_i \mathbf{z}(k)$$

Vectors $\mathbf{f}_i$ can be selected such that they have nonzero elements in the positions corresponding to either inputs or outputs in the $\mathbf{z}(k)$. If we introduce $\tilde{y}(k) = \mathbf{f}_i \mathbf{z}(k)$ or $\tilde{u}(k) = \mathbf{f}_i \mathbf{z}(k)$, this type of penalty can be regarded as a penalization on filtered variables where the filter has an FIR character. The criterion can contain more than one type of filter, in fact any linear combination of these filters can be used without difficulty. If $\tilde{\mathbf{z}}(k)$ represents our pseudo-state made up of delayed inputs and outputs, we can use criterion

$$J = \sum_{t_0+1}^{t_0+T} \tilde{\mathbf{z}}^T(k)\mathbf{Q1}\tilde{\mathbf{z}}(k) + \tilde{\mathbf{z}}^T(k)\mathbf{Q2}\tilde{\mathbf{z}}(k) + \ldots \tilde{\mathbf{z}}^T(k)\mathbf{Qn}\tilde{\mathbf{z}}(k) \qquad (6.41)$$

Designing and choosing individual filters seems simpler than selecting an entire matrix so, in our example, we have chosen to penalize the output difference because this is where the oscillation of the response is most apparent. The filter used was the simple $f(z) = [1 - z^{-1}]$, the vector of which was $\mathbf{f}_1 = [0\ldots0\ 1 - 1\ldots0]$, resulting in criterion matrix

$$\mathbf{Q} = \begin{bmatrix} Q_u \ldots & 0 & \ldots & 0\ 0 \\ 0 \ldots & 0 & \ldots & 0\ 0 \\ 0 \ldots & Q_y + Q_{dy} & -Q_{dy}\ldots & 0\ 0 \\ 0 \ldots & -Q_{dy} & Q_{dy} & 0\ 0 \\ 0 \ldots & 0 & \ldots & 0\ 0 \\ 0 \ldots & 0 & \ldots & 0\ 0 \end{bmatrix}$$

Q_u, Q_y, Q_{dy} are the parameters used to modify the criterion. The LQ toolbox contains the procedure which is used to apply these modifications, as well as to penalize the increments on input $u(k) - u(k-1)$. The structure of this procedure can serve as a model for developing other types of penalization.

Modifying the Open-Loop Transfer function

The dynamic penalization creates a greater space for tuning control behaviour to suit the user but does not guarantee that all conditions are satisfied. A typical example is the requirement for zero steady state error in the step response. We have demonstrated that one approach is to add an integrator factor into the open-loop transfer function. This can be generalized in that it is possible to introduce any transfer function into the loop. Synthesis for a new system must then be performed, consisting of a serial combination of the original systems plus the new transfer functions. The algorithmic solution is straightforward. If we wish to include an integrator we proceed as follows:

1. Before starting the optimization process we create a new denominator for the system

$$\bar{a} = (1 - z^{-1})A(z^{-1})$$

By doing this we state that there will be an integrator in the loop.

2. Once optimisation has been completed we expand controller polynomial R in a similar fashion. This action actually adds the integrator to the loop.

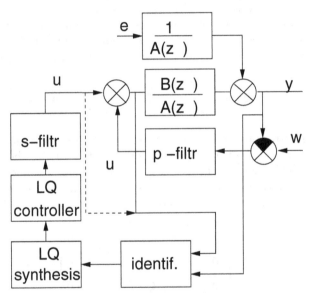

Figure 6.58. A flow diagram of an adaptive LQ controller with parallel and serial filters

If a general transfer function is added to the system, the system's numerator and denominator must be modified in the same way. R and S in the controller must also be altered after optimization.

This approach can be combined with LQ or any of the classic designs for correction terms. If the filter design results in control behaviour which approaches optimal, the synthesis will suggest an LQ controller which only corrects the loop slightly. We can take this to extremes. If we design a filter to be LQ optimal, the controller transfer function will equal one.

We have given an example of adjusting an LQ controller in series with a correcting filter. A similar approach can be taken towards designing an LQ controller working in parallel with a filter. Here the controller must be designed for a system which already has another controller $R1$ in the feedback path. The design here is not based on the original system, but on a system with transfer function

$$S_n = \frac{S}{1 + SR1}$$

This is a common method which is often unwittingly used in controlling processes containing internal feedback. It is a standard approach to unstable processes where the stabilizing feedback is found first, allowing a further controller to implement the required loop characteristics.

The advantage of this connection is that, when Q_u is used as a tuning device, the closed loop transfer function is changed from the primary controller to the optimal LQ loop. In addition to this, parallel connection creates a natural back up. The disadvantage is a more complicated recalculation of the system during the LQ design process. This drawback is overcome when an adaptive form of controller is used, since the consideration of different signals for input during identification can be used in identifying either the system itself or a closed loop connection of the system with a primary controller.

Generally, two LQ controllers can be connected in parallel and it has been found that adaptive versions work reliably.

Cross-product Term in the Criterion

Until now we have considered penalizations of u and $\mathbf{x}$ strictly separated. However, the criterion (6.41) makes it possible to set a cross-product term $(u^T \mathbf{Q}_{ux} \mathbf{x})$. In the literature the cross-term is usually not considered or it is shown that it can be eliminated by substitution [125]

$$\tilde{u} = u + Q_u^{-1} \mathbf{Q}_{ux} \mathbf{x} \tag{6.42}$$

Here we will use the cross-term product penalization $\mathbf{Q}_{ux}$ to tune an LQ controller to give interesting and useful properties. Let us consider penalization matrix

$$\mathbf{Q}_A = \alpha \begin{bmatrix} 1 & \mathbf{L}_A \\ \mathbf{L}_A^T & \mathbf{L}_A^T \mathbf{L}_A \end{bmatrix} \tag{6.43}$$

then the minimization of the criterion

$$J = \sum_{t_0+1}^{t_0+T} \tilde{\mathbf{z}}(t) \alpha \mathbf{Q}_A \tilde{\mathbf{z}}(t) \tag{6.44}$$

leads to a control law $u^*(t) = \mathbf{L}_A \mathbf{x}(t-1)$. $\mathbf{L}_A$ is any fixed linear controller including any PID type. Using the criterion

$$J = \sum_{t_0+1}^{t_0+T} \tilde{\mathbf{z}}(t) (\mathbf{Q} + \alpha \mathbf{Q}_A) \tilde{\mathbf{z}}(t) \tag{6.45}$$

where Q is any other penalization matrix constructed according to (6.41) giving reasonably tuned LQ controller. Then by varying value α from ∞ to 0 we can tune the LQ controller from any fixed controller with the control law $\mathbf{L}_A$ to a standard LQ controller.

Note.

When using criterion (6.45) and the control law $\mathbf{L}_A$ is not stabilizing, the resulting control law from the optimization will be stabilizing even for large but finite values of α. When $\mathbf{L}_A$ is stabilizing, then for large α the controller resulting from the optimisation will be close to $\mathbf{L}_A$.

The use of reference signal $u0$

The input reference signal $u0$ can be used in controller settings in various ways.

1. Naturally, it has the meaning of the input value characterizing the operating point of the plant.

2. It can be considered as another variable to be optimized during the optimization process. Typically, its value can be determined so that the steady state step response has no offset [124].

3. It need not be a constant value, it can be variable over time, $u0(t)$. Let it be a function (linear) of the output and delayed inputs and output. We can write

$$u0(t) = \mathbf{L}_A x(t-1) \tag{6.46}$$

so $u0(t)$ is generated by some controller. As the LQ controller output consists of several parts (6.5) the resulting controller is a weighted parallel connection of a controller $\mathbf{L}_A$ with an LQ controller. The weight is Q_u . For $Q_u \to 0$ the influence of $\mathbf{L}_A$ is negligible, for $Q_u \to \infty$ the LQ controller follows the signal $u0$. In this way, by changing Q_u, different LQ controller properties can be generated. Typically, this property can be used for cooperation of existing standard controllers.

Note. As the LQ optimisation does not know how $u0$ is generated the stability of such a connection cannot be guaranteed.

Variable Penalizations.

Adaptive environment enables the use of varying penalization; time, data or environment dependent. The optimization is solved in each sampling period because parameters of the plant model can change. But there is a reason to assume that the criterion can change as well. The change of criterion can be influenced by the user. Such a possibility was outline above. Two other possibilities are treated here, examples of data-dependent penalizations.

Penalization of input depending on input saturation

In one step strategy (horizon $T = 1$) or in the last step of a multi-step strategy it is easy to design an additional input penalization so that the input at that time is just on the boundary of the allowed interval.

$$Q_u(u_t, u_{sat}) \begin{cases} Q_u = \frac{u_t - u_{sat}}{H_{uu}}, u_t > u_{sat} \\ Q_u = qu, u_t \le u_{sat} \end{cases} \tag{6.47}$$

Q_u is the new penalization, $\mathbf{H}_{uu}$ is part of the Riccati matrix (see Section 6.7).
Penalization of output depending on the amplitude of the disturbance.
Similarly, it is possible to create a controller that compensates only disturbances that exceed a given boundary. Then the variable penalization is Q_y
and when the disturbance is below the specified boundary, it is set to zero. So

$$Q_y(y_t, y_0, \varepsilon) = \begin{cases} Q_y, if(abs(y_t) > y_0) \\ 0, otherwise \end{cases} \qquad (6.48)$$

Q_y can also depend on the amplitude of the disturbance, hence

$$Q_y(y_t, y_0, \varepsilon) = \begin{cases} \alpha Q_y(abs(y_t) - y_0), if(abs(y_t) > y_0) \\ 0, otherwise \end{cases} \qquad (6.49)$$

6.5.2 Implementing Linear Quadratic Controllers

There is a whole range of problems to be solved before applying LQ adaptive
controllers to a chosen system, certainly more and of greater complexity than
for the classic types of controller. The result may be (significantly) improved
control quality and an ability to meet specific technological requirements, fast
step response with small (or even without) overshoot, for example.

It therefore makes sense to try to apply an LQ controller in those cases
where a classic controller, however well adjusted, cannot provide the required
quality of control, even though we suspect there is room for improvement. No
controller can increase the step response speed if the system input is saturated
even for PI controller. We can hardly expect recursive identification to provide
a good model for the synthesis of optimal control if the measured variables
are heavily affected by noise, contain significant drift or systematic error.

Once we decide to apply an LQ controller, the following preparations must
be made:

1. We must decide which model to use (structure, order). This involves determining input and output, and perhaps additional measured variables,
   and the number of delayed terms to be considered in the model regressor
   (order).
2. The size of the forgetting factor must be selected according to the nature
   of the changes in the parameters.
3. We must check that the parameter estimates of the regression model result
   in unbiased estimates.
4. We must decide whether to use an incremental model (include the integrator component in the control loop).
5. Generate the start-up conditions for recursive identification using *a priori*
   knowledge.

The process of connecting up and tuning the controller can be speeded up by using a computer program to detect problems and provide the means to solve them before the controller is actually activated. The ABET programming system for MATLAB® [126, 127], is one such program and can perform the following operations:

1. Collect and filter measured data on the process.
2. Determine the structure on the basis of the data measured, together with the ability to include *a priori* knowledge of the process in various forms.
3. Identify parameters from the measured data. There is a choice of several forms and levels of forgetting. We can evaluate the authenticity of the parameters obtained, and therefore the entire model, from the resulting covariance matrix of the parameters.
4. Design a controller for the model we have obtained and test it by simulation. Provided the model is sufficiently precise, this module can be used to adjust and test the controller so that transfer to a real plant is as smooth as possible. If a model obtained from identification and used for synthesis is also used in simulation, the testing can only be limited because the complexity and nonlinearity of the system will not be represented.

Introducing the Controller into the Control Loop

The synthesis process is reliable and represents the deterministic transformation of model to controller parameters. This transformation is modified by the criterion. The one uncertain variable is the length of the horizon, which must be selected so to obtain stabilizing control.

There are three reasons why an adaptive controller might fail:

1. The system model was ill-chosen
2. The identification lead to parameters which do not accurately reflect the controlled process, causing insufficient robustness and unacceptable behaviour.
3. The horizon of the criterion is small

The Start

Usually an adaptive controller is activated so that the process is controlled by another controller (or manually) and the adaptive controller is switched on at a given moment. Bumpless conditions must be ensured at the moment of switching so that the data register in the adaptive controller receives the true values of signals measured previously. This can be done by allowing the controller to act as an observer over a certain period of time. It can thus measure individual variables, perform identification and synthesis, but is not allowed to apply the results. At a suitable moment, when the process is at its quietest, the system input is switched to the adaptive controller output. The result can be broken down into three categories:

1. Behaviour is good or acceptable. In this case all is well and fine tuning can be carried out by changing the penalty, provided the application permits this.
2. After the switch is made the system reacts with a large, unwanted excitation, but the process settls down again in time and control behaviour is then good. This is a typical reaction to badly chosen initial estimates for the model parameters and can be avoided by improving the estimates. This can be done by increasing the length of the identification period, using additional data for identification, and by supplying whatever *a priori* knowledge is available. The fact that the process returns to normal and the controller functions satisfactorily shows that the structure of the process model is adequate and that identification provides parameter estimates which give a good representation of the system.
3. The worst case is where control is so poor that the adaptive controller must be disconnected. This may be caused by:
   (a) bad model structure (order too low) unable to represent the real process;
   (b) the identification process giving biased parameter estimates, perhaps due to the violation of conditions of an unbiased estimate by the disturbance;
   (c) the criterion chosen could not produce a controller with sufficient robustness;
   (d) the process is markedly nonlinear and cannot be represented by linear model.

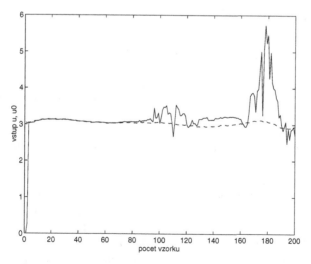

Figure 6.59. The successive transfer of control from PID to LQ controller (u full, $u0$ dashed

When this occurs we can try to increase the order of the model, thus creating more freedom for identification to respect the properties of the disturbance and the system. We can also modify the criterion to result in a more robust controller. In the previous section we saw that increasing penalty Q_u and lengthening the sampling period improve robustness, though this is accompanied by a deterioration in the quality which can be achieved. Nonlinearity in the process is usually such that it is impossible to obtain a more precise linear model. Neither raising the order nor more thorough identification will lead to any improvement. The only thing that can be done is to try to improve the robustness of the controller. The tuning process is not a one-shot event and often the whole procedure must be repeated several times.

We have already shown how to use input reference signal $u0$ in combining an LQ adaptive controller with another type. As can be seen from Figure 6.59, the selection of just one parameter suffices to move from a reliable standard controller to a fully operational adaptive controller. This figure illustrates the input of a controlled system, which was a heat exchanger to make hot service water [128]. The output of a standard PID controller was used as signal $u0$. In the initial stage of adaptation, where the adaptive controller is apt to give inappropriate controller output, a high value was assigned to penalty Q_u This resulted in the adaptive controller, in effect, following signal $u0$ so that the process was, to all intents and purposes, controlled in the standard way. Gradually, penalty Q_u was decreased to levels suitable for LQ control. The penalty was chosen on the basis of earlier experimentation. The LQ control algorithm progressively gained in effectiveness as the penalty was lowered.

6.6 Multivariable Control

The multivariable version of LQ control, that is where there are several system inputs and outputs, creates no complications either theoretically or algorithmically. Individual variable vectors will appear in regression model (6.4) and criterion (6.2). LQ control is solved in the state space, where the vectors and matrices are also used for systems with one input and output. Equally, there are few algorithmic difficulties to be solved for multivariable examples, other than the extra dimensions of the vectors and matrices and the consequent growth in the number of operations and increased calculation time.

There is, however, a complication in the analysis of the behaviour of multivariable controller and how it is tuned. In our one-dimensional example penalty $Q_y = 1$ was sufficient, but in a multivariable process we must deal with the $\mathbf{Q}_y$ matrix and, at the same time, it is difficult to give precise instructions. An indication of the procedure is given in [25]. On the other hand, multivariable controllers have been applied in practical experiments [129] and some applications are described in Section 8.3.

Unless the nature of the process requires a truly multivariable approach, it is more sensible to approximate this need using a series of one-dimensional controllers with external disturbance created by the components left over from the output. A fundamental difference is that each output is directly determined by one input. Yet the controller output is also determined by the other outputs. This modification has one advantage: the sampling period for the controller used in diferent loops can differ. Closer analysis and programming support falls outside the scope of this chapter.

6.7 Minimization of the Quadratic Criterion

In this section we will be dealing with a detailed approach to minimizing the quadratic criterion using the principle of dynamic programming, both in standard form and in a square root form, which will form the basis for our own calculation algorithm.

6.7.1 Standard Minimization of the Quadratic Criterion

We will take the form of quadratic criterion

$$J = \sum_{i=k+1}^{k+T} \left[q_y \left(w\left(i \right) - y\left(i \right) \right)^2 + q_u \left(u\left(i \right) - u0\left(i \right) \right)^2 \right] + \mathbf{x}^T(T)\mathbf{S}_0\mathbf{x}(T) \quad (6.50)$$

which is the most generalized form appearing in this chapter. When minimizing criterion (6.2) or (6.50), it is formally appropriate to use the state form to describe the system. This need not be a state space model in the classic sense of the term, the state space form alone is sufficient. The condition is that the vector on the left side of the equation in time k is a function of the same vector in time $k-1$. If the form of regression vector (6.12) is taken the following notation is obtained:

$$\mathbf{x}\left(k \right) = \mathbf{P}_x\mathbf{x}\left(k-1 \right) + \mathbf{P}_u u\left(k \right) + \mathbf{P}_v\bar{\mathbf{v}}\left(k-1 \right) = \mathbf{P}\mathbf{z}\left(k-1 \right) \quad (6.51)$$

where

$$\mathbf{x}^T\left(k \right) = \left[u\left(k \right), u\left(k-1 \right), \ldots, u\left(k-nb \right), y\left(k \right), y\left(k-1 \right), \ldots, y\left(k-na+1 \right) \right]$$
$$\mathbf{z}^T\left(k \right) = \left[u\left(k \right), \mathbf{x}\left(k-1 \right), \bar{\mathbf{v}}\left(k-1 \right) \right]$$
$$\bar{\mathbf{v}}\left(k-1 \right) = \left[v\left(k-1 \right), v\left(k-2 \right), \ldots, v\left(k-nd \right) \right]$$

It is assumed that disturbance $v(k)$ is generated by filter $1/Av(z^{-1})$. The individual matrices are obtained from the definitions of the vectors

$$\mathbf{P}_x = \begin{bmatrix} 1 & 0 & 0 & \cdots & 0 & 0 \\ 0 & \cdots & 0 & \cdots & 0 & 0 \\ b_1 & \cdots & b_{nb} & a_1 & \cdots & a_{na} \\ 0 & \cdots & 0 & 1 & 0 & \cdots \\ 0 & \cdots & 0 & \cdots & 0 & 0 \\ 0 & \cdots & 0 & \cdots & 0 & 0 \\ 0 & \cdots & 0 & \cdots & 0 & 0 \end{bmatrix} ; \mathbf{P}_u = \begin{bmatrix} 1 \\ 0 \\ b_0 \\ 0 \\ \cdots \\ 0 \end{bmatrix}$$

$$\mathbf{P}_v = \begin{bmatrix} 0 & 0 & 0 & 0 \\ 0 & 0 & 0 & 0 \\ d_1 & \cdots & d_{nd} & k \\ 0 & 0 & 0 & 0 \\ 0 & 0 & 0 & 0 \\ av_1 & \cdots & av_{nd} & 0 \\ 1 & 0 & 0 & 0 \end{bmatrix}$$

State matrix $\mathbf{P}_x$ also respects any autoregression model of the development of measurable noise. Matrix $\mathbf{P} = [\mathbf{P}_u, \mathbf{P}_x, \mathbf{P}_v]$ is composed of a row containing the parameters of the regression model, the parameters for the models of the development of external disturbance (av_i) and the appropriate number of 1s. Vector $\tilde{\mathbf{z}}(k)$ is now used, which also contains other variables upon which the value of the criterion depends. So quadratic criterion (6.2) can be written

$$J = \sum_{i=k_0+1}^{k_0+T} \tilde{\mathbf{z}}^T(k)\, \mathbf{Q} \tilde{\mathbf{z}}(k) \tag{6.52}$$

where $\tilde{\mathbf{z}}^T(k) = [\mathbf{x}(k), w(k), u0(k)]$, and the standard penalty is drawn from

$$\mathbf{Q} = \begin{bmatrix} Q_u & \cdots & 0 & \cdots & 0 & -Q_u \\ 0 & \cdots & 0 & \cdots & 0 & 0 \\ 0 & \cdots & Q_y & \cdots & -Q_y & 0 \\ 0 & \cdots & 0 & \cdots & 0 & 0 \\ 0 & \cdots & -Q_y & \cdots & Q_y & 0 \\ -Q_u & \cdots & 0 & \cdots & 0 & Q_u \end{bmatrix}$$

Generally, $\mathbf{Q}$ can be any symmetrical positive semi-definite matrix.

The minimization process (6.52) now proceeds from the end of the horizon T as follows: $u(k_0 + T)$ can only affect the final term of (6.52). If we express $y(k_0 + T)$ in vector $\mathbf{x}(k_0 + T)$ as a function of the preceding $y(k)$ and $u(k_0 + T)$ we can then find $u(k_0 + T)$ to minimize $y(k_0 + T)$. Matrix $\mathbf{P}$ from (6.51) is used in the new calculation.

Since the vector $\tilde{\mathbf{z}}(k)$ contains further elements which do not change we can write

$$\tilde{\mathbf{z}}(k_0 + T) = \begin{bmatrix} \mathbf{P} & 0 & 0 \\ 0 & 1 & 0 \\ 0 & 0 & 1 \end{bmatrix} \tilde{\mathbf{z}}(k_0 + T - 1) = \mathbf{\Omega}\tilde{\mathbf{z}}(k_0 + T - 1)$$

The size of the losses we wish to minimize using $u(k_0 + T)$ is, therefore

$$\begin{aligned} J(k_0 + T) &= \tilde{\mathbf{z}}^T(k_0 + T - 1)\,\mathbf{\Omega}^T\mathbf{Q}\mathbf{\Omega}\tilde{\mathbf{z}}(k_0 + T - 1) \\ &= \tilde{\mathbf{z}}^T(k_0 + T - 1)\,\mathbf{H}\tilde{\mathbf{z}}(k_0 + T - 1) \end{aligned} \tag{6.53}$$

The minimum of this quadratic form is obtained in the standard way (derivation by $u(k_0 + T)$, with the condition that the derivation equals zero). We obtain

$$\begin{aligned} J^*(k_0 + T) &= \min\left[J(k_0 + T)\right] \tag{6.54} \\ &= \bar{\mathbf{z}}^T(k_0 + T - 1)\left(\mathbf{H}_{xx} - \mathbf{H}_{xx}^T\mathbf{H}_{uu}^{-1}\mathbf{H}_{ux}\right)\bar{\mathbf{z}}(k_0 + T - 1) \end{aligned}$$

where $\bar{\mathbf{z}}(k_0 + T - 1)$ is $\tilde{\mathbf{z}}(k_0 + T - 1)$ without element $u(k_0 + T)$, and $\mathbf{H}_{..}$ are the corresponding submatrices of matrix $\mathbf{H}$. Because $\mathbf{H}$ represents loss in time $k_0 + T$, we coordinate it with this index.

Loss was minimized in the final step so we can now turn to minimizing the penultimate step. This component of the criterion depends on $u(k_0 + T - 1)$ and is

$$\begin{aligned} J^*(k_0 + T) &= \min\left[J(k_0 + T)\right] \\ &= \bar{\mathbf{z}}^T(k_0 + T - 1)\mathbf{H}_{xx}^*\bar{\mathbf{z}}(k_0 + T - 1) + \\ &\quad\ \tilde{\mathbf{z}}^T(k_0 + T - 1)\mathbf{Q}\tilde{\mathbf{z}}(k_0 + T - 1) \end{aligned} \tag{6.55}$$

where matrix $\mathbf{H}_{xx}^* = (\mathbf{H}_{xx} - \mathbf{H}_{ux}^T\mathbf{H}_{uu}^{-1}\mathbf{H}_{ux})$ remains from the previous step. Vectors $\tilde{\mathbf{z}}(k_0 + T - 1)$ and $\bar{\mathbf{z}}(k_0 + T - 1)$ differ in that vector $\tilde{\mathbf{z}}(k_0 + T - 1)$ contains set point $w(k_0 + T - 1)$. We can now create a combined vector $\tilde{\mathbf{z}}(k_0 + T - 1)$, which contains both final set points $w(k_0 + T)$ and also $w(k_0 + T - 1)$. We must also expand $\mathbf{H}^*$ and $\mathbf{Q}$ by a column of zeros at the appropriate position. We can now continue formally as in the preceding step. We express $y(k_0 + T + 1)]$ in vector $\mathbf{x}(k_0 + T - 1)$ as a function of the previous $y(k)$ and $u(k_0 + T - 1)$. Matrix $\mathbf{\Omega}$ will now be given

$$\Omega = \begin{bmatrix} \mathbf{P}\ 0\ 0\ 0 \\ 0\ 1\ 0\ 0 \\ 0\ 0\ 1\ 0 \\ 0\ 0\ 0\ 1 \end{bmatrix}$$

We can see that matrix Ω has acquired an extra column which corresponds to variable $w(k_0 + T - 1)$, which will continue to appear until the end of minimization. The loss in the penultimate step will be

$$J(k_0 + T - 1) = \tilde{\mathbf{z}}^T(k_0 + T - 1)\,\Omega^T\,(\mathbf{H}^*(k_0 + T) + \mathbf{Q})\,\tilde{\mathbf{z}}(k_0 + T - 1)$$
$$= \tilde{\mathbf{z}}^T(k_0 + T - 1)\,\mathbf{H}(k_0 + T - 1)\,\tilde{\mathbf{z}}(k_0 + T - 1)$$

(6.56)

The result of minimization is similar to the previous step.

$$J^*(k_0 + T - 1) = \min_{u_0 + T - 1}\,[J(k_0 + T - 1)] \tag{6.57}$$
$$= \bar{\mathbf{z}}^T(k_0 + T - 2)\left(\mathbf{H}_{xx} - \mathbf{H}_{xx}^T\mathbf{H}_{uu}^{-1}\mathbf{H}_{ux}\right)\bar{\mathbf{z}}(k_0 + T - 2)$$

This procedure is continued until all the terms of the criterion have been processed.

Commentary

We have shown that the control law can be split into several parts according to which variable is used to multiply their elements. The basic component is $\mathbf{H}_{ux}$, which is multiplied by variables u and y. This component defines the feedback properties and represents controller transfer function $\frac{u(z)}{y(z)} = \frac{S(z^{-1})}{R(z^{-1})}$. The relationship between polynomials S and R and vector $\mathbf{H}_{ux}$ is given by the relation

$$R = \begin{bmatrix} 1 & \mathbf{H}_{ux}\,(1\ :\ nb) \end{bmatrix}; \qquad S = \begin{bmatrix} 0 & \mathbf{H}_{ux}\,(nb + 1\ :\ nb + na) \end{bmatrix}$$

The $\mathbf{H}_{uv}$ part represents the feedforward (filter) of external disturbance. There may be several of these components, depending on how many external disturbances have been modelled. Part $\mathbf{H}_{uw}$ represents the influence of the set point. We have seen how each value for $w(k)$ is given by single element $\mathbf{H}_{uw}$. If the criterion contains many terms (a long horizon), the dimension of $\mathbf{H}_{uw}$ will be large. It is, however, possible to accumulate these values in one column. If we assume that the set point does not change along the horizon, then all the elements of vector $\mathbf{H}_{uw}$ can simply be added and the result used as the numerator of the controller transfer function from set point $F_{uw} = \frac{\mathbf{H}_{uw}}{R}$. Since variable $w(k)$ can be regarded as the product $w(k) * 1$ values can be included in matrix Ω. Vector $\tilde{\mathbf{z}}(k)$ will then contain 1s for $w(k)$. Any set point change can then be accumulated in one column using this layout. Similarly, $\mathbf{H}_{uu0}$ represents the influence of the reference input variables and the same rules apply as for $\mathbf{H}_{uw}$.

6.7.2 Minimization of the Quadratic Criterion in Square Root Form

The procedure given above is unsuitable for practical calculations. The difference which appears in relation (6.54) can cause the matrix of the quadratic form $\mathbf{H}_{xx}$ to cease to be positive definite (semi- definite). This means that the calculated value no longer makes any sense. The so-called square root approach, where we work with the square root (factor) of the symmetrical matrix, is a useful tool for avoiding this. All the algorithms in this part are based on this principle.

The square root of a matrix is defined as follows. Matrix $\mathbf{F}$ is the square root of symmetrical, positive definite matrix $\mathbf{A}$, when

$$\mathbf{A} = \mathbf{F}^T \mathbf{F}$$

is valid. $\mathbf{F}^T$ is the matrix transposed to $\mathbf{F}$. There is an infinite number of matrices which satisfy the condition above. However, there exists an unambiguous factorization on a triangular matrix. Therefore

$$\mathbf{A} = \mathbf{\Delta}^T \mathbf{\Delta} = \mathbf{\nabla}^T \mathbf{\nabla}$$

yet $\mathbf{\Delta}^T \neq \mathbf{\nabla}$, whie $\mathbf{\Delta}$ and $\mathbf{\nabla}$ mark the lower and upper triangular matrix respectively.

The guiding principle of the square root is to replace the symmetrical matrix by its square root (factor of the matrix) for the purposes of calculation. In the minimization of criterion (6.52) we used matrix multiplication, addition of matrices, and the relation for minimizing quadratic form (6.54). In the same way, when we take the square root approach, we will require specific operations to manipulate the square root matrix.

Matrix multiplication remains, but we will need to define the operations which correspond to the addition of the matrices and minimization of the quadratic criterion. It soon becomes clear that the factor of the matrix addition corresponds to the expansion of the matrix of one factor by another. If $\mathbf{A} = \mathbf{F}^T \mathbf{F}$, $\mathbf{B} = \mathbf{G}^T \mathbf{G}$, and $\mathbf{C} = \mathbf{H}^T \mathbf{H}$, then it follows from the definition of the multiplication of the composed matrices that

$$\mathbf{C} = \mathbf{A} + \mathbf{B} = \mathbf{H}^T \mathbf{H} = [\mathbf{FG}] [\mathbf{FG}]^T$$

therefore

$$\mathbf{H} = \begin{bmatrix} \mathbf{F} \\ \mathbf{G} \end{bmatrix}$$

Minimization of the quadratic form: according to relation (6.54), minimization of the quadratic form is used to find the $u(k)$ which minimizes

$$[u(k), \mathbf{x}(k-1)] \mathbf{S} [u(k), \mathbf{x}(k-1)]^T \tag{6.58}$$

If we use factor $\mathbf{F}$ instead of $\mathbf{S}$, (6.58) can be written as

$$\left\| \mathbf{F} \left[u\left(k\right), \mathbf{x}\left(k-1\right) \right]^{T} \right\|^{2}$$

If $\mathbf{F}$ is the upper triangle, $u(k)$ can only influence the value of the first element of the vector under consideration. We obtain minimization by selecting

$$\mathbf{F}_{uu} u\left(k\right) + \mathbf{F}_{ux} \mathbf{x}\left(k-1\right) = 0$$

where $\mathbf{F}_{uu}$ and $\mathbf{F}_{ux}$ are submatrices of $\mathbf{F}$ corresponding to the multiplication by $u(k)$ or $\mathbf{x}(k-1)$. The minimum achieved is given by the relation

$$\left\| \mathbf{F}_{xx} \mathbf{x}\left(k-1\right) \right\|^{2}$$

In the square root approach a further operation must be added to those above. Both the multiplication operation and the expansion of the matrix disturb the triangular shape of the square root of the quadratic form. Yet we still need the triangular shape both for its inambiguity and in order to find the optimal $u(k)$. The operation we require is orthogonal transformation, which can restore the triangular form. Orthogonal transformation is represented by regular square matrix $\mathbf{T}$, where $\mathbf{T}^{-1} = \mathbf{T}^{T}$, $i.e.$ $\mathbf{T}^{T}\mathbf{T} = \mathbf{I}$. In reality, the multiplication of the matrix square root using the orthogonal matrix from the left does not change the value of the quadratic form.

$$\mathbf{x}^{T}\mathbf{S}\mathbf{x} = \mathbf{x}^{T}\mathbf{F}^{T}\mathbf{F}\mathbf{x} = \mathbf{x}^{T}\mathbf{F}^{T}\mathbf{T}^{T}\mathbf{T}\mathbf{F}\mathbf{x} = \mathbf{x}^{T}\mathbf{\Delta}^{T}\mathbf{\Delta}\mathbf{x}$$

Algorithms for the $\mathbf{T}$ transformation, which transforms $\mathbf{F}$ into $\mathbf{\Delta}$, are known under the names Householder transformation, Givenson transformation, or elementary rotation.

6.7.3 The Minimization Algorithm

The minimization algorithm comprises the above relations with the proviso that:

- Instead of using the square root matrix, it operates with factorization of the symmetrical matrix on $\mathbf{S} = \mathbf{U}^{T}\mathbf{D}\mathbf{U}$ where $\mathbf{U}$ is the triangular matrix with a unit diagonal, and $\mathbf{D}$ is the diagonal matrix.
- The orthogonal transformation is realized by the MATLAB® qr function or employs the method of elementary rotations coded in special function RTREDUC.

The algorithm can be described as follows:
1. Create matrix $\mathbf{S}$ with dimension (nx, nz) with diagonal $\mathbf{S}_0$. $\mathbf{S}_0 = 1000$ is standard. The algorithm is not sensitive to this value. Set $i = horizon$
2. Calculate mean value $\mathbf{H} = \mathbf{S}\mathbf{P}$;
3. Augment $\mathbf{H}$ with all penalizations

$$\tilde{\mathbf{H}} = \begin{bmatrix} \mathbf{H} \\ \alpha_1 \mathbf{f}_1 \\ \ldots \\ \alpha_2 \mathbf{f}_2 \end{bmatrix}$$

4. Perform orthogonal transformation of $\tilde{\mathbf{H}}$ into triangle matrix $\mathbf{H}_\Delta$.
5. Split matrix $\mathbf{H}_\Delta$ into submatrices as follows

$$\begin{bmatrix} \mathbf{H}_{uu} & \mathbf{H}_{ux} \\ 0 & \mathbf{H}_{xx} \end{bmatrix}$$

6. Do:$i = i - 1$; $if(i > 0)$, $\mathbf{S} = \mathbf{H}_{xx}$; goto step2
7. Control law is

$$\mathbf{cl} = -\mathbf{H}_{uu}^{-1}\mathbf{H}_{ux}$$

Graphically the process can be represented by Figure 6.60

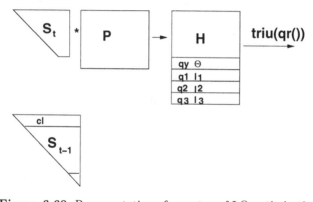

Figure 6.60. Representation of one step of LQ optimization

The operations described are performed at each step of the minimization of the quadratic criterion (one iteration of the Riccati equation).

The form of vectors $\mathbf{l}_i$ representing the quare root of the most widely used penalizations are presented in the Table 6.5.

6.8 Summary of chapter

LQ adaptive control can offer nice closed loop behaviour for uncertain plant when

- correct model can be identified;

Table 6.5. Basic penalizations

Penalizations of	Weight	$u(k)$	$u(k-1)$	...	$y(k)$	$y(k-1)$	$y0$	$u0$
input	α_u	1	0	0	0	0	0	-1
output	α_y	b_0	b_1	...	a_1	a_2	-1	0
input increment	$\alpha_{\Delta u}$	1	-1	0	0	0	n 0	0
output increment	$\alpha_{\Delta y}$	b_0	b_1	...	$a_1 - 1$	$a_2 - 1$	0	0

- the system is sufficiently linear.

Unfortunately, these conditions can be checked only *ex post*, when some trial is done. That is why operators are very cautious when using something that might not work properly. In addition, the choice of penalties to obtain satisfactory behaviour is not simple. To change the situation the following was done: properties of an LQ controller in disturbance compensation, set point tracking and the frequency domain were documented Tuning was simplified by factorization of the penalty matrix to a sum of rank one matrices having specific physical meaning. LQ controller tuning was enriched by the possibility to emulate any standard controller and by the change of one parameter (penalization) smoothly changing the controller behaviour from the standard controller to an LQ one, and vice versa if necessary. Reliability is ensured by using square root approaches to both system on-line identification and controller synthesis.

Problems

6.1. Compensation of stochastic disturbance.
(a) Simulate a continuous-time plant with stochastic disturbance and discrete adaptive controller. Use a disturbance which is correctly modelled by the identified regression model. Observe how the quality of controldepends on the penalization Q_u and how the input amplitude depends on Q_u.
(b) On the same scheme with Q_u fixed, change the sampling period of the controller (and identification as well) and observe disturbance compensation and input signal amplitude.
(c) Filter the disturbance so that disturbance dynamics do not correspond to plant dynamics. Observe the identified parameters. What could help (filtering of signals for identification, increasing the order of the model)?

6.2. Set point control and offset removal.
(a) Observe how the offset value depends on input penalization Q_u or in-

crement of input penalization $Q_{\Delta u}$. Use instead an adaptive controller with integrator. (b) Demonstrate how the offset is removed when using a reference input signal with $u0 = 1/g$ $y0$ ($g=$ plant static gain).
(c) Compare the step response when future values of the set point can or cannot be used. Observe the role of the horizon in the criterion.

6.3. Startup of the adaptive algorithm.
(a) Observe the role of prior information for startup of the adaptive algorithm. Demonstrate that even correct prior parameter estimates are not sufficient without reasonable choice of parameter covariance matrix and initial data (initial inputs and outputs).
(b) Create a SIMULINK® scheme in which the adaptive controller first collects data, several sample times later starts the identification, and after some further tine switches to the adaptive LQ controller. During the first two phases use a simple fixed controller.
(c) Connect the output of the selected fixed controller tothe $u0$ port of the adaptive LQ controller. Observe the smooth transition between LQ behaviour and simple controller behaviour when changing Q_u. Design appropriate changes of Q_u to obtain bumpless startup.

6.4. Generalized penalizations.
(a) Verify on an example (e.g. model no. 3. or the helicopter) that for systems with slightly damped poles, nonzero penalization of $Q_{\Delta y}$ is necessary to obtain a step response without overshoot.
(b) Design a penalization which will make the LQ controller imitate the behaviour of an I (PI, PID, any fixed linear controller described by a transfer function) controller. Observe that the resulting control law is the one that you want to imitate;it is stabilizing.
(c) Observe, that if the choice of alternative controller leads to an unstable closed loop, optimization will not result in such a control law.

6.5. Forgetting in identification.
Using the helicopter nonlinear model verify the role ofthe forgetting factor.

Computer-aided Design for Self-tuning Controllers

In the education process of Automatic Control Theory, the verification through simulation, and the practical implementation of the designed controller algorithms in real-time conditions, are very important for training control engineers. In Chapter 7 therefore, two MATLAB® toolboxes are presented which contain digital self-tuning controllers that have been described in the previous chapters of this monograph. Section 7.1 describes the Self-tuning Controllers SIMULINK® Library (STCSL), where digital PID controllers (see Chapter 4) and controllers based on the algebraic approach (see Chapter 5) are included. The LQ toolbox, containing self-tuning controllers based on general regression model identification and LQ criterion optimization, and presented in Chapter 6, is described in Section 7.2.

7.1 Self-tuning Controllers Simulink® Library

Based on Chapter 4 and Chapter 5, a library of self-tuning controllers was created in the MATLAB®/SIMULINK® environment [130], the purpose being to create an environment suitable for the creation and testing of self-tuning controllers. This library is available free of charge from the Tomas Bata University in Zlín Internet site www.utb.cz/stctool [104]. The library was created using MATLAB® version 6.0 (Release 12), but it can be ported, with some changes, to both lower and higher MATLAB® versions. Controllers are implemented in the library as standalone SIMULINK® blocks, which allow easy incorporation into existing simulation schemes and simple creation of new simulation circuits. Only standard techniques of the SIMULINK® environment were used when creating the controller blocks and so only a basic knowledge of this environment is required to be able to begin working with the library. Controllers can be implemented into simulation schemes just by copying or using the drag & drop function, and their parameters are set using dialogue windows. Another advantage of the approach used is its relatively easy implementation of user-defined controllers by the modification of some suitable controller in the library.

Nowadays, the library contains over 30 simple single-input/single-output, digital, self-tuning controllers, which use discrete second- and third-order models for recursive process identification. All of these controllers use discrete control laws, in which controller parameters are computed by various methods. The first part covers self-tuning algorithms using the traditional PID structure; the second part describes controllers based on the algebraic approach. The library package not only contains the controllers, but also includes a reference manual with simple descriptions of the algorithm and the internal structure of each controller. The controllers included in the library use the discrete ARX model for the recursive identification of process model parameters. The library user can select one of three recursive identification methods offered for the computation of parameter estimates vector $\hat{\Theta}(k)$: the basic least squares method or one of its modifications, the least squares method with exponential forgetting and the least squares method with adaptive directional forgetting (see Chapter 3).

The typical wiring of any library controller is shown in Figure 7.1. Each self-tuning controller from the library uses three input signals and provides two outputs. The inputs are the reference signal (w) and the actual output of the controlled process (y). The last controller input is the current input of the controlled process – the control signal (u_in). The value of this signal does not have to be the same as the controller output, $e.g.$ due to controller output saturation. The main controller output is, of course, the input signal of the controlled process. The second controller output consists of the current parameter estimates of the controlled process model. The number of parameters this output consists of depends on the model used by on-line identification: controllers using the second-order model provide four values, $i.e.$ (a_1, a_2, b_1, b_2); and controllers using the third-order model provide six values, $i.e.$ $(a_1, a_2, a_3, b_1, b_2, b_3)$.

A scheme analogous to the scheme in Figure 7.1 can also be used to simulate the control processes of both discrete and continuous-time controlled processes with much more complicated structures. It is also possible to implement processes with time variable parameters, processes described by nonlinear differential equations, $etc.$

7.1.1 Overview of Library Controllers

The Self-tuning Controllers SIMULINK® Library is started by opening the following SIMULINK® file: $stcsl\_std.mdl$, which contains block schemes of all controllers.

The name of the controller always corresponds to the name of the file that processes the calculation of the controller parameters and always has the following structure: **xxNyyyy**. The first two characters (**xx**), indicate

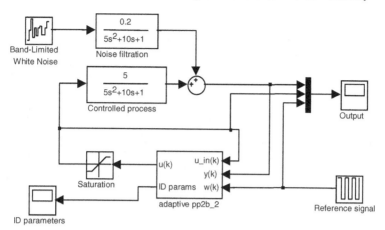

Figure 7.1. Control circuit in the SIMULINK® environment

the controller type – **zn** indicates a controller synthesis based on the Ziegler–Nichols method; **pp** represents controllers with a pole assignment synthesis; **mv** denotes the minimum variance controller, *etc.* The third character (**N**) in a controller name is the digit 2 or 3 – corresponding to the order of the model used by the on-line identification part of the controller. The following characters (**yyyy**), serve to cover more detailed controller details. A survey of all the library controllers is set out in Table 7.1.

Table 7.1. Survey of individual controllers

Contr. name	Controller algorithm	Input param.
(1) zn2fr zn3fr	$u(k) = q_0 e(k) + q_1 e(k-1) + q_2 e(k-2) + u(k-1)$ $q_0 = K_P(1 + \frac{T_D}{T_0}); \qquad q_1 = -K_P(1 - \frac{T_0}{T_I} + 2\frac{T_D}{T_0})$ $q_2 = K_P \frac{T_D}{T_0}$	K_{Pu} T_u T_0
(2) zn2br zn3fr	$u(k) = q_0 e(k) + q_1 e(k-1) + q_2 e(k-2) + u(k-1)$ $q_0 = K_P(1 + \frac{T_D}{T_0} + \frac{T_0}{T_I}); \qquad q_1 = -K_P(1 + 2\frac{T_D}{T_0})$ $q_2 = K_P \frac{T_D}{T_0}$	K_{Pu} T_u T_0
(3) zn2br zn3fr	$u(k) = q_0 e(k) + q_1 e(k-1) + q_2 e(k-2) + u(k-1)$ $q_0 = K_P(1 + \frac{T_D}{T_0} + \frac{T_0}{2T_I}); \qquad q_1 = -K_P(1 - \frac{T_0}{2T_I} + \frac{2T_D}{T_0})$ $q_2 = K_P \frac{T_D}{T_0}$	K_{Pu} T_u T_0
(4) zn2pd	$u(k) = K_P \left\{ w(k) - y(k) + \frac{T_D}{T_0}[y(k-1) - y(k)] \right\}$ $K_P = 0.4K_{Pu}; \qquad T_D = \frac{T_U}{20}$	K_{Pu} $T_u; T_0$

(5)	$u(k) = q_0 e(k) + q_1 e(k-1) + u(k-1)$	K_{Pu}
zn2pi	$q_0 = K_P \left(1 + \frac{T_0}{2T_I}\right);$ $q_1 = -K_P \left(1 - \frac{T_0}{2T_I}\right)$	$T_u; T_0$
(6)	$u(k) = p_1 u(k-1) + p_2 u(k-2) + q_0 e(k)$	K_{Pu}
	$+ q_1 e(k-1) + q_2 e(k-2)$	
	$T_f = \frac{T_D}{\alpha};$ $\alpha \in \langle 3; 20 \rangle$	
zn2fd	$p_1 = \frac{-4\frac{T_f}{T_0}}{\frac{2T_f}{T_0}+1};$ $p_2 = \frac{\frac{2T_f}{T_0}-1}{\frac{2T_f}{T_0}+1}$	T_u
zn3fd	$q_0 = \frac{K_P + 2K_P \frac{T_f+T_D}{T_0} + \frac{K_P T_0}{2T_I}(\frac{2T_f}{T_0}+1)}{\frac{2T_f}{T_0}+1}$	T_0
	$q_1 = \frac{\frac{K_P T_0}{2T_I} - 4K_P \frac{T_f+T_D}{T_0}}{\frac{2T_f}{T_0}+1}$	
	$q_2 = \frac{\frac{T_f}{T_0}(2K_P - \frac{K_P T_0}{T_I}) + 2\frac{K_P T_D}{T_0} + \frac{K_P T_0}{2T_I} - K_P}{\frac{2T_f}{T_0}+1}$	
(7)	$u(k) = K_R[y(k-1) - y(k)] + K_I[w(k) - y(k)]$	K_{Pu}
zn2tak	$+ K_D[2y(k-1) - y(k-2) - y(k)] + u(k-1)$	T_u
zn3tak	$K_R = 0.6 K_{Pu} - \frac{K_I}{2};$ $K_I = \frac{1.2 K_{Pu} T_0}{T_u}$	T_0
	$K_D = \frac{3 K_{Pu} T_u}{40 T_0}$	
(8)	$u(k) = u_{PI}(k) + u_D(k)$	K_{Pu}
	$\alpha \in \langle 3; 20 \rangle;$ $\beta \in \langle 0; 1 \rangle$	
zn2ast	$u_{PI}(k) = K_P[y(k-1) - y(k)] + \frac{K_P T_0}{2T_I}[e(k) + e(k-1)]$	T_u
zn3ast	$+ \beta K_P[w(k) - w(k-1)] + u_{PI}(k-1)$	T_0
	$u_D(k) = K_P \frac{T_D \alpha}{T_D + T_0 \alpha}[y(k-1) - y(k)] + \frac{T_D}{T_D + T_0 \alpha} u_D(k-1)$	
(9)	$u(k) = K_P \left\{ e(k) - e(k-1) + \frac{T_0}{T_I} e(k) \right.$	K_{Pu}
zn2fpd	$+ \frac{T_D}{6T_0}[e(k) + 2e(k-1) - 6e(k-2) + 2e(k-3) + e(k-4)] \}$	T_u
zn3fpd	$+ u(k-1)$	T_0
(10)	$u(k) = q_0 e(k) + q_1 e(k-1) + q_2 e(k-2) + u(k-1)$	d
ba2	$\gamma = \frac{b_1}{b_0}$	
	$k_I = \frac{1}{2d-1};$ for $\gamma = 0;$ $k_I = \frac{1}{2d(1+\gamma)(1-\gamma)};$ for $\gamma > 0$	
	$q_0 = \frac{k_I}{b_0};$ $q_1 = q_0 a_1 = \frac{k_I}{b_0} a_1;$ $q_2 = q_0 a_2 = \frac{k_I}{b_0} a_2$	
(11)	$u(k) = K_P \left\{ e(k) - e(k-1) + \frac{T_0}{T_I} e(k) \right.$	B
	$+ \frac{T_D}{T_0}[e(k) - 2e(k-1) + e(k-2)] \} + u(k-1)$	
da2	$Q = 1 - e^{-\frac{T_0}{B}};$ $K_P = -\frac{(a_1 + 2a_2)Q}{b_1}$	T_0
	$T_D = \frac{T_0 a_2 Q}{K_P b_1};$ $T_I = -\frac{T_0}{\frac{1}{a_1 + 2a_2} + 1 + \frac{T_D}{T_0}}$	
(12)	$u(k) = q_0 e(k) + q_1 e(k-1) + q_2 e(k-2) + (1-\gamma)u(k-1)$	ω_n
	$+ \gamma u(k-2)$	
pp2a_1	$s_1 = a_2 \left[(b_1 + b_2)(a_1 b_2 - a_2 b_1) + b_2(b_1 d_2 - b_2 d_1 - b_2) \right]$	

	$r_1 = (b_1 + b_2)(a_1 b_1 b_2 + a_2 b_1^2 + b_2^2$	
	$\gamma = q_2 \frac{b_2}{a_2};$ $q_0 = \frac{1}{b_1}(d_1 + 1 - a_1 - \gamma)$	
	$q_1 = \frac{a_2}{b_2} - q_2(\frac{b_1}{b_2} - \frac{a_1}{a_2} + 1);$ $q_2 = -\frac{s_1}{r_1}$	ξ
(13)	$u(k) = q_0 e(k) + q_1 e(k-1) + q_2 e(k-2) + (1-\gamma)u(k-1)$	
	$+\gamma u(k-2)$	α
pp2a_2	$r_1 = (b_1 + b_2)(a_1 b_1 b_2 + a_2 b_1^2 + b_2^2)$	
	$r_2 = x_1(b_1 + b_2)(a_1 b_2 - a_2 b_1)$	ω
	$r_3 = b_1^2 x_4 - b_2[b_1 x_3 - b_2(x_1 + x_2)]$	
	$r_4 = a_1[b_1^2 x_4 + b_2^2 x_1 - b_1 b_2(x_2 + x_3)]$	
	$r_5 = (b_1 + b_2)[a_2(b_1 x_2 - b_2 x_1) - b_1 x_4 + b_2 x_3]$	
	$r_6 = b_1(b_1^2 x_4 - b_1 b_2 x_3 + b_2^2 x_2) - b_2^3 x_1$	
	$\gamma = \frac{r_6}{r_1};$ $q_0 = \frac{r_2 - r_3}{r_1};$ $q_1 = -\frac{r_4 + r_5}{r_1};$ $q_2 = \frac{x_4 + \gamma a_2}{b_2}$	
(14)	$u(k) = -[(q_0' + \beta)y(k) - (q_0' + q_2')y(k-1) + q_2' y(k-2)]$	
	$-(\gamma - 1)u(k-1) + \gamma u(k-2) + \beta w(k)$	ω_n
pp2b_1	$s_1 = a_2 \left\{ b_2[a_1(b_1 + b_2) + b_1(d_2 - a_2) - b_2(d_1 + 1)] - a_2 b_1^2 \right\}$	ξ
	$r_1 = (b_1 + b_2)(a_1 b_1 b_2 - a_2 b_1^2 - b_2^2)$	
	$q_0' = q_2'(\frac{b_1}{b_2} - \frac{a_1}{a_2}) - \frac{a_2}{b_2};$ $q_2' = \frac{s_1}{r_1};$ $\gamma = q_2' \frac{b_2}{a_2}$	
	$\beta = \frac{1}{b_1}(d_1 + 1 - a_1 - \gamma - b_1 q_0')$	
(15)	$u(k) = -[(q_0' + \beta)y(k) - (q_0' + q_2')y(k-1) + q_2' y(k-2)]$	α
	$-(\gamma - 1)u(k-1) + \gamma u(k-2) + \beta w(k)$	
	$r_1 = (b_1 + b_2)(a_1 b_1 b_2 - a_2 b_1^2 - b_2^2)$	
pp2b_2	$r_2 = a_1 b_2[b_1(x_2 + x_3 + x_4) - b_2 x_1]$	ω
	$r_3 = a_2 b_1[b_2 x_1 - b_1(x_2 - x_3 + x_4)]$	
	$r_4 = (b_1 + b_2)[b_1 x_4 + b_2(-x_3 - x_4)]$	
	$r_5 = b_1(b_1^2 x_4 - b_1 b_2 x_3 + b_2^2 x_2) - b_2^3 x_1$	
	$r_6 = b_1^2(-a_2 x_3 + a_1 x_4 - a_2 x_4)$	
	$r_7 = b_2[b_1(a_1 x_4 + a_2 x_2 - x_4) - b_2(a_2 x_1 + x_4)]$	
	$q_0' = -\frac{r_2 - r_3 + r_4}{r_1};$ $q_2' = \frac{r_6 + r_7}{r_1}$	
	$\gamma = \frac{r_5}{r_1};$ $\beta = \frac{x_1 + x_2 - x_3 + x_4}{b_1 + b_2}$	
(16)	$u(k) = r_0 w(k) - q_0 y(k) - q_1 y(k-1)$	
db2w	$-q_2 y(k-2) - p_1 u(k-1) - p_2 u(k-2)$	
db3w		
(17)	$u(k) = r_0 w(k) - q_0 y(k) - q_1 y(k-1)$	
db2s	$-q_2 y(k-2) - p_1 u(k-1) - p_2 u(k-2)$	
db3s		

(18)	$u(k) = \frac{1}{q}[a_1 y(k-1) + a_2 y(k-2) - b_1 u(k-1)$ $- b_2 u(k-2) + w(k)] + u(k-1)$	q
mv2		
(19)	$u(k) = r_0 w(k) + r_1 w(k-1) + r_2 w(k-2) - q_0 y(k)$ $- q_1 y(k-1) - q_2 y(k-2) - p_1 u(k-1) - p_2 u(k-2)$	D
pp2chp	$A(z^{-1})P(z^{-1}) + B(z^{-1})Q(z^{-1}) = D(z^{-1})$	T_0
pp3chp	$B(z^{-1})R(z^{-1}) + D_w(z^{-1})S(z^{-1}) = D(z^{-1})$	
(20)	$u(k) = r_0 w(k) + r_1 w(k-1) + r_2 w(k-2)$ $- q_0 y(k) - q_1 y(k-1) - q_2 y(k-2) - p_1 u(k-1) - p_2 u(k-2)$	q_u
pp2lq	$A(z^{-1})P(z^{-1}) + B(z^{-1})Q(z^{-1}) = D(z^{-1})$ $B(z^{-1})R(z^{-1}) + D_w(z^{-1})S(z^{-1}) = D(z^{-1})$ $D(z^{-1}) = 1 + d_1 z^{-1} + d_2 z^{-2}$ $m_0 = q_u(1 + a_1^2 + a_2^2) + b_1^2 + b_2^2$ $m_2 = q_u a_2; \qquad m_1 = q_u(a_1 + a_1 a_2) + b_1 b_2$ $d_1 = \frac{m_1}{\delta + m_2}; \qquad d_2 = \frac{m_2}{\delta}; \qquad \delta = \frac{\lambda + \sqrt{\lambda^2 - 4m_2^2}}{2}$ $\lambda = \frac{m_0}{2} - m_2 + \sqrt{\left(\frac{m_0}{2} + m_2\right)^2 - m_1^2}$	T_0

Notes

The undermentioned relations are used for computiation of the individual controller parameters.

Controller numbers: 1–4, 6, 8 and 9:

$$K_P = 0.6 K_{Pu} \qquad T_I = 0.5 T_u \qquad T_D = 0.125 T_u$$

Controller numbers 12 and 14:

$$d_1 = -2\exp(-\xi \omega_n T_0)\cos(\omega_n T_0 \sqrt{1 - \xi^2}) \quad \text{for } \xi \le 1$$

$$d_1 = -2\exp(-\xi \omega_n T_0)\cosh(\omega_n T_0 \sqrt{1 - \xi^2}) \quad \text{for } \xi > 1$$

Controller numbers 13 and 15:

$$x_1 = c + 1 - a_1 \qquad x_2 = d + a_1 - a_2 \qquad x_3 = f + a_2; \qquad x_4 = g$$

$$c = -4\alpha \qquad d = 6\alpha^2 + \omega^2 \qquad f = -2\alpha(2\alpha^2 + \omega^2) \qquad g = \alpha^2(\alpha^2 + \omega^2)$$

7.1.2 Controller Parameters

The dynamic behaviour of a controller can be influenced by changing the parameters available for a given controller. Controller parameters can be divided into two groups:

- parameters common to all controllers;
- controller specific parameters.

All controller parameters are set or changed by standard approaches in the SIMULINK® environment, that is, by changing items in the dialogue window – invoked for example, by double-clicking with the mouse on the controller object. This dialogue window also contains a short description of the corresponding controller; and the "Help" button is used to invoke the corresponding part of the hypertext reference guide. The parameter setting dialogue window of the *pp2a_1* controller is shown in Figure 7.2.

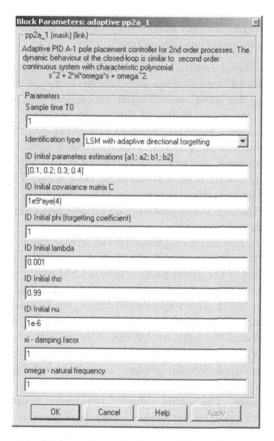

Figure 7.2. Dialogue for *pp2a_1* controller setting parameters

The group of common controller parameters consists of the sampling period and parameters affecting the on-line identification process. The list of these parameters is set out here.

Sample time T0. It defines sampling period T_0. This value is used for process identification as well as for calculation of the controller output.

Identification type. It is possible to select from the following on-line identification methods:

- Least squares method (LSM) – the simplest method, in which all process input/output pairs affect identified parameters with the same weight.
- LSM with exponential forgetting – the latest process input/output pairs affect identified parameters more than older pairs. This method can be used for systems with time-varying parameters.
- LSM with adaptive directional forgetting – the most sophisticated method, the weight of the current process input/output pair is determined with respect to changes of input and output signal. This method is useful for systems with time-varying parameters.

ID Initial parameter estimations. The initial process parameter estimates used by on-line identification. This is a column vector of parameters in the form $[a_1, a_2, b_1, b_2]$ for most second-order controllers; $[a_1, a_2, b_1]$ for Dahlin controller; and $[a_1, a_2, a_3, b_1, b_2, b_3]$ for third-order systems.

ID Initial covariance matrix C. Initial value of the covariance matrix used in the on-line identification process. This must be a square positive definite matrix with dimensions the same as the number of identified parameters. Usually, a diagonal matrix is used in the form: $\mathbf{G} * \mathbf{I}$, where $\mathbf{I}$ is the unit matrix and $\mathbf{G}$ is gain. The gain then determines the influence of initial parameter estimates on the identification process: the greater the gain, the smaller the influence of the initial estimates.

ID Initial phi (forgetting coefficient). The initial value of the forgetting factor φ used in the on-line identification of the controller process. This parameter is used only if the *Identification type* is LSM with exponential forgetting or LSM with adaptive directional forgetting and should be in the range: $0 < \varphi \leq 1$. When LSM with exponential forgetting is used, this parameter determines the forgetting rate of the older process input/output pairs: the smaller φ, the smaller the influence of the older pairs to the current parameter estimates. When LSM with adaptive directional forgetting is used, φ changes during the identification process, according to the process of input u and output y values.

ID Initial lambda. The initial value of parameter λ used in the on-line identification of the controlled process. This parameter is used only if the *Identification type* is LSM with adaptive directional forgetting.

ID Initial rho. The initial value of parameter ρ used in the on-line identification of the controlled process. This parameter is used only if the *Identification type* is LSM with adaptive directional forgetting.

ID Initial nu. The initial value of parameter ν used in the on-line identification of the controlled process. This parameter is used only if the *Identification type* is LSM with adaptive directional forgetting.

Some controllers allow adjustment by parameters specific to the particular controller – these controller-specific parameters are listed in Table 7.2. For example, *pp2a_1* controller-specific parameters are *xi – damping factor* and *omega – natural frequency*, as shown in Figure 7.2.

Table 7.2. Controller-specific parameters

Type	Parameter	Description
ba2	ID dead time	The dead time d of the controlled process during sampling periods. The value of this parameter must be a non-negative integer.
da2	B – adjustment factor	The adjustment factor B specifies the dominant time constant of the closed loop: the smaller B, the quicker the closed loop step response.
pp2a_1 *pp2b_1*	xi – damping factor	The damping factor ξ specifying the dynamic behaviour of the closed loop. The dynamic behaviour of the closed loop is similar to that of second-order continuous-time systems with a characteristic polynomial $D(s) = s^2 + 2\xi\omega_n s + \omega_n^2$.
	omega – natural frequency	The natural frequency ω_n specifying the dynamic behaviour of the closed loop. The dynamic behaviour of the closed loop is similar to second-order continuous-time systems with characteristic polynomial $D(s) = s^2 + 2\xi\omega_n s + \omega_n^2$.
pp2a_2 *pp2b_2*	omega – imaginary component of the pole	The imaginary component ω of the poles of the closed loop. The dynamic behaviour of the closed loop is defined by its poles: $z_{1,2} = \alpha \pm j\omega$; $z_{3,4} = \alpha$.
	alfa – real component of the pole	The real component α of the poles of the closed loop. The dynamic behaviour of the closed loop is defined by its poles: $z_{1,2} = \alpha \pm j\omega$; $z_{3,4} = \alpha$.
zn2fd *zn3fd*	alfa – filtration coefficient	The filtration coefficient α used to filter the process output signal. The time constant of the filter is $T_f = T_D/\alpha$, usually $\alpha \in \langle 3; 20 \rangle$.
mv2	q – penalisation factor	The penalisation factor q is used for the computation of control law parameters. This parameter specifies the measure of change of the current controller output with respect to the previous controller output: the smaller the penalisation, the greater the possible change of controller output.

zn2ast *zn3ast*	alfa – filter constant	The filtration coefficient α is used to filter the process output signal. The time constant of the filter is, $T_f = T_D/\alpha$, usually $\alpha \in \langle 3; 20 \rangle$.
	beta – weight factor	The weight β of the reference signal in the proportional component of the controller. This weight factor should fulfil the condition: $0 < \beta \leq 1$.
pp2chp *pp3chp*	Reference signal type	The controller is capable of working with three types of controlled signal: steps, ramps and sine waves.
	Frequency	The frequency of sine waves of the reference signal. This parameter is used only if reference signal type is sine waves.
	Coefficients of characteristic polynomial	The vector of coefficients of the closed loop characteristic polynomial $D(z^{-1}) = d_0 + d_1 z^{-1} + d_2 z^{-2} + \ldots$. The vector is formed as $[d_0, d_1, d_2, \ldots]$.
pp2lq	Reference signal type	The controller is capable of working with three types of controlled signal: steps, ramps and sine waves.
	Frequency	The frequency of sine waves of the reference signal. This parameter is used only if reference signal type is sine waves.
	fi – penalization of controller output	The penalization of controller output in the LQ criterion. The higher the value of q_u, the greater the weight of controller output in the LQ criterion.

7.1.3 Internal Controller Structure

Each library controller is constructed as a mask of a subsystem, which consists of SIMULINK® blocks and has inputs, outputs and parameters. As stated in the previous chapters, each controller uses three input signals (*i.e.* the reference signal w, the controlled signal y, and the control signal u_in), and provides two outputs (*i.e.* the control output u, and the current model parameter estimates).

The internal controller structure consists of SIMULINK® blocks which provide, among other things, for the possibility of simple creation of a new controller by the modification of an appropriate library controller. The structure of the controller *pp2a_1* is presented in Figure 7.3 as an example. Each library controller consists of three basic parts:

- on-line identification block;

- block for computing controller parameters;
- block for computing controller output.

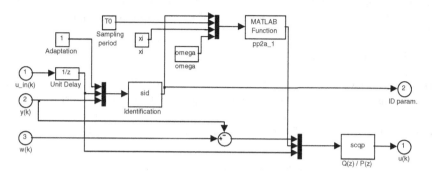

Figure 7.3. Block scheme of the *pp2a_1* controller

From a programmer point of view, each block corresponds to a stand-alone program file, which is, after the simulation starts, interpreted by the MATLAB® environment. The blocks are implemented as a MATLAB® *Fcn* and an *S-Function*. The MATLAB® *Fcn* block performs the call of a MATLAB® function, which converts the input data vector to the output data vector. The *S-Function* block is more universal, and the corresponding function can in addition use both discrete and continuous-time states.

The on-line identification algorithm is implemented using the s-function *sid*, which calculates parameter estimations of the controlled process model on the basis of the current process output, the previous process input, the identification parameters, and the states of the on-line identification.

The computation of controller parameters is provided by the m-function, which does not use the states, and only converts the input vector to the output vector. The input vector consists of the current parameter model estimates, and eventually, other parameters depending on controller type. The output vector contains the controller parameters used when calculating the controller output. The name of this function is the same as the name of the whole controller.

The computation of controller output is implemented as an s-function computing the controller output on the basis of the controller parameters, the control error, and the previous input of controlled output. The reference value and the current process output are used instead of control error in feedback feedforward controllers. The previous value of the controller output is not taken as a state – because it can be changed before reaching the controlled process input, *e.g.* due to saturation. The *scqp* function is used to calculate controller output in most of the feedback controllers, *i.e.* controllers that use control error. Controllers containing a feedforward part, use the *scrqp* function. Controllers using four-point difference (*zn2fpd* and *zn3fpd*) and Åström

Controllers (*zn2ast* and *zn3ast*) use their own functions to compute controller output – *scfpd* and *scast*.

The basic version of the library uses the MATLAB® programming language to implement all of the above-mentioned functions. The advantage of this approach is the simple implementation of matrix and vector operations.

7.1.4 Reference Guide and Help

Besides the files implementing the functionality of all library controllers, the library package also includes a reference guide with a short description of all the controller algorithms, operating with the controllers, and a description of the internal structure of the controllers. The guide is currently available in two versions:

- *pdf* format – suitable for printing;
- *html* format – used for context help.

Moreover, the help for each function included in the library can be invoked by entering the following: *help function_name* in the MATLAB® command line. Then a short description of the function, its inputs, and outputs is displayed in the MATLAB® command window.

7.1.5 Creating Applications with Real-Time Workshop

The MATLAB®/SIMULINK® environment can also be used to generate code to be used in controllers in industrial practice. *Real-Time Workshop*, one of the toolboxes shipped with MATLAB®, allows the generation of source code and programs to be used outside the MATLAB® environment. The process of generating the source code is controlled by special compiler files that are interpreted by *Target Language Compiler*. These files are identified by the *.tlc* (target language compiler) extension and describe how to convert SIMULINK® schemes into the target language. Thereby, source code is generated and after compiling and linking, the resulting application is created. Applications for various microprocessors and operating systems can be created by selecting the corresponding *.tlc* files.

The *Target Language Compiler* can create applications to be used under the Windows environment, which perform control algorithms and save the results in a binary file with a structure acceptable to the MATLAB® system. An analysis of the control process can then be performed using the advantages of the MATLAB® functions and commands. Selecting another *.tlc* file, leads to the creation of an MS-DOS application, or an application to be used on PC-based industrial computers without the requirement for an operating system.

Many manufacturers of industrial computers and controllers have created their own target language compiler files used to create applications for the equipment they produce. *Real-Time Workshop* provides a relatively open environment for the conversion of block schemes to various platforms, where any users can create their own *.tlc* files for conversion of the block scheme to a source code and hence, achieve compatibility with any hardware.

Application creation only requires selection of the appropriate *.tlc* file, or eventually, setting the compiler parameters and then initiation of the compilation process.

Compatibility of Simulink® Schemes

The Target language compiler files provided in a standard installation of *Real-Time Workshop* unfortunately do not support all SIMULINK® blocks. The set of unsupported blocks depends on the *.tlc* file used, but when working with the Self-tuning Controllers SIMULINK® Library, the user may encounter problems especially with two blocks: the MATLAB® *Fcn* and *S-Function*. The *.tlc* files do not support the MATLAB® *Fcn* block, and the *S-Function* block is supported only in cases where the function is written in the C language without calls in MATLAB®.

The standard library version *stcsl_std* has been revised, and the *stcsl_rtw* version created so as to achieve full compatibility of the library and *Real-Time Workshop*. The functionality and internal structure of all controllers is the same in both versions, the only difference is in the resources used to program the individual parts of the controllers. The MATLAB® Fcn blocks are used in the standard version to compute controller parameters. These blocks have been superseded by C language s-functions, which only compute their outputs (controller parameters) on the basis of their inputs (parameter estimates of the controlled process model and eventually, other controller-specific parameters) – these functions do not use any states. The on-line identification block and the blocks computing controller output are implemented as s-functions that have had to be rewritten into C language. The file names in both library versions are the same – the difference is in the extension: functions written in MATLAB® language use an *.m* extension, and functions written in C language use a *.c* extension.

Library Versions

As mentioned in the previous chapter, the Self-tuning Controllers SIMULINK® Library is available in two versions: the standard version and the *Real-Time Workshop* version. The functionality, number of controllers, and internal controller structure is the same in both versions – the difference being in the internal program implementation of the program code

The standard version uses MATLAB® programming language, with a simple code structure, where no variable declarations are required – the program interpreter assigns the type to the variable according to its usage. Another advantage of this language is the very simple implementation of matrix and vector operations. The on-line identification block especially uses matrix multiplications and transpositions and the computation of dead-beat controller parameters requires matrix inversion. The code written in this language is easy to read, to study, and to understand – and thus suitable when studying the details of controller implementation. This code is also suitable for writing user-defined controllers.

The *Real-Time Workshop* version of the library was constructed using the C programming language. The main advantage of this version is its portability, because only functions written in C language are supported when converting block schemes to applications through the *Target Language Compiler*. The disadvantage resides in larger source files and thus worse readability of the code.

The Self-tuning Controllers SIMULINK® Library is used in the university course Adaptive Control Systems. Its architecture enables easy user orientation in SIMULINK® block schemes and generates source code for controller functions. The controllers provided are suitable for modification and thereby implementation of user-defined controllers. Compatibility with *Real-Time Workshop* not only ensures the possibility of laboratory testing using real-time models, but also the possibility of creating applications for industrial controllers.

7.2 Linear-Quadratic Toolbox

The toolbox given here has arisen out of the need to demonstrate the characteristics and application of

- LQ controllers, which require knowledge of the discrete transfer function of the controlled system.
- Adaptive LQ controllers.

The current toolbox for MATLAB® and SIMULINK® provides:

- SIMULINK® schemes for:
  - a fixed controller (various schemes to demonstrate various characteristics).
  - an adaptive controller (a standard regression model, or a model with integrator factor).
- SIMULINK® blocks representing fixed and adaptive controllers.
- m-files for LQ optimization and identification and a few auxiliary m-files.

7.2.1 Fixed Linear-Quadratic Controllers

This part of the LQ toolbox was created to provide the material presented in Chapter 6 demonstrating the properties of this LQ approach. There are two basic schemes. *Schema1.mdl* is depicted in Figure 6.4, and serves for discrete time applications. The *Schema2.mdl* demonstrates the application of a fixed controller to a continuous-time plant.

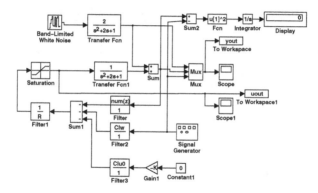

Figure 7.4. *Schema2.mdl*; discrete LQ controller with continuous-time plant

This scheme is similar to that of the general controller scheme Figure 6.2. The input signals are system output $y(t)$; set point $w(t)$; and reference input $u0(t)$. The parameters of such a controller can be calculated using of the function

   function [R,S,kon,Clw,Clu0,sric]=lqexb(b,a,k,nstep,qur,dqy,dqu,s0)

where the outputs of the function are the parameters of corresponding blocks, and additionally, the square root of the Riccati matrix ($sric$); input parameters are as follows:

b - System numerator in the operator z^{-1} with arbitrary length, *i.e* leading zeros can be used to represent time delay.

a - System denominator of arbitrary order.

k - Value of the offset in a regression model (6.4).

$nstep$ - Horizon of the criterion.

qur - Penalization of system input.

dqy - Penalization of system output increments.

dqu - Penalization of system input increments.

$s0$ - Diagonal element of the initial Riccati matrix

   The code of this function is quite simple.

   The auxiliary function: *abk2pr.m* creates a pseudostate matrix of the system; *makesm1.m* creates the initial Riccati matrix. The function *znob.m* contains the call of *lqexb.m*. This function is useful for tuning the controller. After changes to any penalization, a new control law is obtained by calling *znob.m*.

7.2.2 Adaptive Controllers

Adaptive controllers are represented by eight SIMULINK® schemes with six different adaptive controllers for the positional LQ adaptive controller (standard regression model) and an incremental one (with added integrator factor) for cases with external measurable disturbance and set point preprogramming.

Listing 7.1. The code of the LQ optimization function *lqexb*

```
function [rr,ss,kon,clw,clu0,s]=lqexb(b,a,k,nstep,qur,dqy,dqu,s0)
    na=max(size(a))-1;
    nb=max(size(b));
    m=nb+na + 3;
    [pr,str]=abk2pr(b,a,k);
    s = makesm1(str,s0,1);
    qur=sqrt(qur); dqu=sqrt(dqu); dqy=sqrt(dqy);
    th=pr(nb,:);
    qu=zeros(size(th)); qdy=th; qdu=qu; qu(1)=1; qu(m)=-1;
    qdy(nb+2)=qdy(nb+2)-1;qdu(1)=1;qdu(2)=-1;
    h=s*pr;
    [mm,nn]= size(h);
    hn= zeros(mm+1,nn); dn=eye(mm+1);
    for i=1:nstep
        h= s*pr;
        xx=th;xx(m-1)=-1;
        hn=triu(qr([h;xx;qur*qu;dqy*qdy;dqu*qdu]));
        s=hn(2:m,2:m);
    end
    cl=hn(1,:)/hn(1,1);
    rr=cl(1,1:nb); ss=cl(1,nb:na +nb); kon=cl(1,m-2); ss(1)=0;
    clw=cl(1,m-1); clu0=cl(1,m);
```

All controllers that do not consider pre-programming use the same basic blocs for identification (*jbslidex.m*), controller synthesis (*lqjbse.m*) and for con-

Table 7.3. Adaptive LQ controller SIMULINK® blocks

Name	Function
adlqs	basic scheme for positional controller
adlqsd	scheme for measurable disturbance and positional controller
adlqsw	scheme for set point pre-programming positional controller
adlqswd	scheme for set point pre-programming
	measurable disturbance with positional controller
adlqi	basic scheme for incremental controller
adlqid	scheme for measurable disturbance and incremental controller
adlqiw	scheme for set point pre-programming incremental controller
adlqiwd	scheme for set point pre-programming
	measurable disturbance with incremental controller

troller (*ctrlx.m*)but differs by the use of other blocks and additional signals. Controllers for pre-programming use slightly modifier functions *ctrlxw* and *lqjbsebw.m*. The SIMULINK® block for identification can be substituted by the identification block *lsident2.mdl* realizing square root identification function *lssqrt.m* having the following code:

Listing 7.2. The code of the square root identification function *lssqrt.m*

```
function [sys, x0,str,ts]= lssqrt(t,x,u,flag,stru,F0,Ts,fi)
          m=stru(2)*stru(1)+stru(4)*stru(3)+stru(6)*stru(5)+1;
  mm=stru(3)*m;
  ny=stru(3);
  nu=stru(1);
       mp=m-1;if (stru(7)==1),mp=m;end
if abs(flag) == 0
       sys = [0 mm+m+(m+ny)*(m+ny)+ny mm+ny  ny+nu 0 0 1];
       x0 = zeros(m+mm+(m+ny)*(m+ny)+ny,1);
       x0(m+1:m+(m+ny)*(m+ny))= F0(:,:);
        if (stru(7)==1),  x0(m)=1; end
       ts =[Ts 0];
    elseif flag == 2
       yy= x(m-ny:m-1);
       uu=u(1:stru(1));
       vv=[ ]; if(stru(5)~=0),vv= u(stru(1)+stru(3)+1:end);end
       xx=x(1:m);
       xx= fun_shif(xx,stru,yy,uu,vv);
       P=zeros(m+ny);
       P(:)= x(m+1:m+(m+ny)*(m+ny));
       z=[xx; u(stru(1)+1:stru(1)+stru(3))];
       P=triu(qr([P;z']));
       Eth=inv(P(1:mp,1:mp))*P(1:mp,m+1:m+ny);
       Eth=[Eth ;zeros(1-stru(7),stru(3))];
       P = P(1:m+ny,1:m+ny)*fi;
       ep =   u(stru(1) +1:stru(3)+stru(1)) -  Eth'*xx ;
         xx(m-ny:m-1)=u(stru(1) +1:stru(3)+stru(1));
       pom=Eth';
       sys=[xx;P(:);pom(:);ep];
    elseif flag == 3
          sys =x(m+(m+ny)*(m+ny) +1:end);
    else
      sys=[ ];
    end
```

The scheme below shows the application of the controller with measurable disturbance and pre-programming in a typical environment.

Controller output is a manipulated variable, the controller block inputs are the system input $u(t)$ (it can be different from the controller output); system output $y(t)$; set point $w(t)$; reference input $u0(t)$; measurable disturbance $d(t)$;

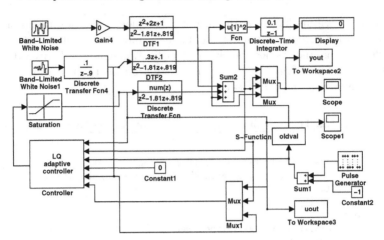

Figure 7.5. *adlqswd.mdl*; adaptive LQ controller with external disturbance and pre-programming

and the pair $[uf(t), yf(t)]$ for identification. These values can be directly: $[u(t), y(t)]$, but frequently they are filtered input and output.

Adaptive controllers are masked blocks. The meaning of its parameters is given in Table 7.4

Table 7.4. Adaptive controller parameters

name	meaning	details
strav	measurable disturbance structure	order of the denominator of the disturbance autoregression model
stru	model structure	[numerator order, denominator order order of the measurable disturbance
qur	input penalization	value of Q_u
dqu	input increment penalization	value of $Q_{\Delta u}$
dqy	output increment penalization	value of $Q_{\Delta y}$
nstep	horizon of the criterion	
fi	forgetting factor	
To	sampling period	

The schemes serve as:

- Demo file.
- A typical simulation environment to test adaptive LQ ControllersIN which the user can change the system or controller setting.
- Sources for the "LQ Adaptive Controller" block for other simulation schemes created by the user. The adaptive controller is provided with a group of S-functions for identification, optimization, and controller output generation.

The blocks of the adaptive controller with exteral disturbance and pre-programming are shown in Figure 7.6. There are two identification blocks providing the parameters of the plant model and of the disturbance model.

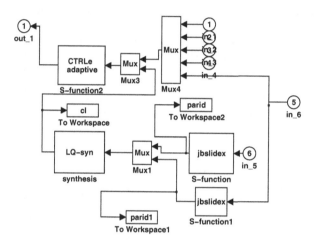

Figure 7.6. Adaptive controller subsystem with external disturbance and pre-programming

The incremental controller is represented by the scheme Figure 7.7.

The parameters of adaptive controllers are set in the mask in Figure 7.8.

7.3 Summary of chapter

Two MATLAB® toolboxes are introduced in this chapter: the Self-tuning Controllers SIMULINK® Library and the LQ Toolbox. The first contains over 30 simple, discrete, single-input/single-output adaptive controllers which identify a controlled process model using second or third-order models. The controller parameters and algorithms are discussed and summarized in overviews in several tables. Incorporation of the controllers into SIMULINK® schemes, simulation verifications and real-time examples are also presented. To simplify

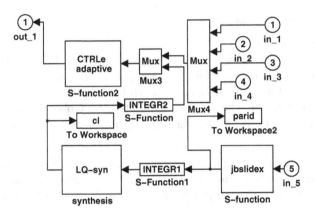

Figure 7.7. Adaptive LQ incremental controller subsystem

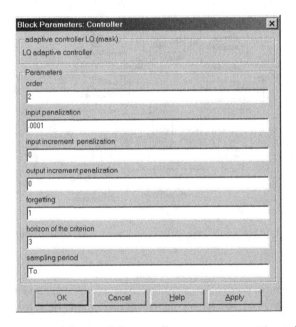

Figure 7.8. Adaptive LQ controller parameters setting table

use of the controllers in real-time environments, the connection to Real-Time Workshop is mentioned. The LQ toolbox contains two groups of discrete controllers based on minimization of the LQ criterion: fixed controllers and adaptive controllers with on-line identification of the controlled process using the ARX model. The control laws of these types of controllers are based on solving the Riccati equation, and this process is presented by listing the program in connection with the corresponding simulation schemes.

8

Application of Self-tuning Controllers

Several self-tuning controllers described in this monograph have been verified by real-time control of laboratory plants. Section 8.1 gives results of the real-time control of several laboratory models by controllers included in the Self-tuning Controllers Simulink Library [131]. Also described are applications of the adaptive LQ controller to a heat exchanger station (Section 8.2), to boiler control with multivariable LQ controllers (Section 8.3), to an adaptive LQ controller in cascade control (Section 8.4) and to steam pressure control in a drum boiler in a power plant (Section 8.5).

8.1 Decentralized Control Using Self-tuning Controllers

The selected SISO controllers included in the library were verified by real-time control of educational laboratory models. This section gives results of the control of the following models: air heating system, laboratory model of coupled motors CE 108, and twin rotor MIMO system (helicopter model). All these models are two-input/two-output (TITO) systems and the decentralized approach with logical supervisor was applied for their control.

The classical approach to the control of MIMO systems is based on the design of a matrix controller to control all system outputs at one time. Computation of the matrix controller is realized by one central computer. The basic advantage of this approach is the possibility of achieving optimal control performance because the controller can use all information known about the controlled system. The disadvantage of using a central matrix controller is its demand on computer resources because the number of operations and required memory depend on the square of the number of controlled signals. Nowadays this problem is reduced thanks to great progress in the development of computer hardware; this, however, increases the price of the control system. Another disadvantage is the influence central controller faults on the controlled system. If the central controller fails, all the controlled signals are affected; thus the reliability of the controller is fundamental. Ensuring the required reliability can be unbearable from the financial point of view, especially in critical applications.

An alternative solution to the control of MIMO systems is a decentralized approach. In this case, the system is considered as a set of interconnected subsystems and the output of each subsystem is influenced not only by the input to this subsystem but also by the input to the other subsystems. Each subsystem is controlled by a stand-alone controller. Thus, decentralized control is based on decomposition of the MIMO system to subsystems, and the design of a controller for each subsystem [104]. Another advantage of the decentralized approach is that it is a lot easier to set controller parameters (*e.g.* choice of poles of the characteristic polynomial) for SISO control loops than for MIMO control loops. On the other hand, the control performance of a decentralized control system is suboptimal because controllers do not use information from the other subsystems. A further disadvantage is the limited applicability of the decentralized control to symmetric systems (systems with an equal number of inputs and outputs).

Each output of a multivariable controlled system can be affected by each system input. The strength of the effect is determined not only by internal transfers of the MIMO system but also by the evolution of the system input signals. When the decentralized approach is used to control such a system then, from the point of view of a controller of a particular subsystem, the transfer function varies in time even if the MIMO system is linear and stable.

The presence of subsystem interconnections is the main reason for using self-tuning controllers in a decentralized approach to ensure the required course of controlled variables. Identification algorithms suitable for use in decentralized control must include weighting of identification data such that new data affect model parameters estimation more than older data. This requirement is a consequence of the time-varying influences of other subsystems on the identified subsystem. The influence of control variable (u_i) on the corresponding controlled variable (y_i) decreases with increasing gain of subsystem interconnections. This could lead to an unstable process of recursive parameter estimates. The stability of recursive identification can be increased by ensuring that just one of the controllers connected to the multivariable systems works in an adaptive regime at a particular time. Recursive identification parts of other controllers are suspended and parameter model estimates are constant for that time.

8.1.1 Supervisory System

The process of switching on and off recursive identification is controlled by a new part of the control circuit – the *supervisory system*. Switching the identification on and off can be described as a process of transferring tokens among subsystems where only the controller, which currently has a token, can perform recursive identification. The token is moved to an other subsystem when a selected criterion is used.

The supervisory system represents a second level of control and thus a control circuit with supervisory system has a hierarchical control structure. An example of a control circuit scheme with supervisory system is shown in Figure 8.1. The first (lowest) level of hierarchy contains individual self-tuning controllers (STC 1, STC 2 and STC 3 in Figure 8.1) and the second level (superior) is represented by a supervisory system, which controls individual self-tuning controllers. The supervisory system analyses particular values from the control circuit and on the basis of these analyses moves the identification token among subsystems. In the case shown in Figure 8.1, the analysis is performed on the basis of reference values and controlled values (dotted lines) and the process of transferring token is represented by dashed lines.

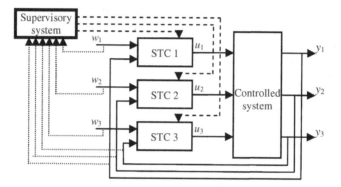

Figure 8.1. Decentralized control circuit with supervisory system

The inclusion of a supervisory system into the control circuit brings the problem of defining a strategy for switching the identification of individual subsystems on and off, *i.e.* moving the token between subsystems.

8.1.2 Critera Used for Ending Adaptation of a Particular Subsystem

Three basic approaches can be used in deciding when to suspend identification of a particular subsystem and move the token to an other one [132]:

- on the basis of elapsed time;
- on the basis of values from the currently identified subsystem;
- on the basis of values from other subsystems.

It is also possible to combine these approaches.

Switching on the Basis of Time
This is the simplest switching algorithm, in which the supervisory system

contains almost no mathematical relations. Switching is realized after a pre-set period that a controller has been in adaptive mode has elapsed. Thus, the controller works in adaptive mode, and after a specific time elapses the parameter estimates are "frozen" and controller continues in a mode with constant parameters and another controllers starts its adaptive mode.

The advantage of this approach is first its simplicity because no values have to be the supervisory system and the programming of switching algorithms is simple. The main disadvantage lies in the fact that the adaptive mode is ended even if the system has not yet been well identified. This could lead to nonstability of the identification process especially if the preset time is short.

Switching on the Basis of Parameters of the Current Subsystem

Particular values from the currently identified subsystem are analysed and when these values fulfil a criterion, the identification is considered accurate enough and the corresponding self-tuning controller is switched from adaptive mode to constant parameters mode. Analysed values can be current estimations of model parameters, the control error, or the controlled value.

If a control variable is observed, the criterion for ending identification is a steady measured level of this variable. The standard deviation computed on the basis of a preset number of previous variable samples can serve as the measure of a steady level. The adaptation mode is ended when the following criterion is fulfilled

$$J_i < eps \qquad J_i = \frac{1}{m}\sqrt{\sum_{j=0}^{m-1}[y_i(k-j)-\bar{y}_i]^2} \qquad \bar{y}_i = \frac{1}{m}\sum_{j=0}^{m-1}y_i(k-j) \quad (8.1)$$

where eps is a threshold decision whether the observed output variable y_i is steady or not, m is the preset number of previous values to be used when computing the standard deviation, and k is the current identification step.

The use of the Equations (8.1) is restricted to cases where the reference signal is constant or sequentially constant, $i.e.$ it contains only step changes. If the reference signal does not fulfil this restriction and thus the controlled signal is not required to trace the sequentially constant signal, the control error can be used instead of the controlled variable

$$J_i < eps \qquad J_i = \frac{1}{m}\sqrt{\sum_{j=0}^{m-1}[e_i(k-j)-\bar{e}_i]^2} \qquad \bar{e}_i = \frac{1}{m}\sum_{j=0}^{m-1}e_i(k-j) \quad (8.2)$$

An advantage of using control error lies in the fact that identification can be considered accurate even in cases where the control error is constant but nonzero. This can occur when the controller is designed for a different reference signal then the one actually used. For example if a pole assignment controller is designed to cover step changes of the reference signal and the reference signal contains a ramp, the control error is constant but nonzero.

Another criterion can be reaching a reference variable, or to be exact, a preset zone width about the reference signal. In this case, the adaptive mode is ended when the control error is less than a predefined level for a predefined number of samples

$$J_i < eps \quad J_i = \max |e_i(k-j)| \quad \text{for } j = 0, 1, \ldots, m \tag{8.3}$$

where eps is a threshold for the decision whether to end the adaptive mode or not, e_i is the control error of the observed subsystem, m is the preset number of previous values to be used when evaluating the criterion and k is the current identification step.

Another approach is based on analysing the ongoing values of model parameter estimates. In this case, the adaptation mode is ended when parameter estimates become steady. The criterion can again be based on the standard deviation of each parameter. Because the system model has more then one parameter, the criterion must cover the ongoing values of all parameters. The maximum standard deviations can be used

$$J_i < eps \quad J_i = \max_r \frac{1}{m} \sqrt{\sum_{j=0}^{m-1} [\Theta_{ir}(k-j) - \bar{\Theta}_{ir}]^2} \quad \bar{\Theta}_{ir} = \frac{1}{m} \sum_{j=0}^{m-1} \Theta_{ir}(k-j) \tag{8.4}$$

where eps is a threshold for the decision when to end the adaptive mode, m is the preset number of previous values to be used when evaluating the criterion, k is the current identification step and $\Theta_{ir}(k-j)$ represents the value of the r-th model parameter of the currently identified subsystem (i-th subsystem).

The advantage of observing parameter estimates, system output or control error is good use of time and thus quicker attainment of correct parameter estimates for all subsystems. There is no waste of identification time by identifying a subsystem which has already been well identified. From the time point of view, observing parameter estimates instead of controlled system output or control error is better because parameter estimates become steady earlier than the system output and movement of the identification token to another subsystem is performed as soon as the parameter estimates become steady. The main disadvantage of this approach lies in higher memory and computer time demands.

Switching on the Basis of Parameters of the Other Subsystems
Ending the adaptive mode of a controller on the basis of parameters of the other subsystems is an opposite approach to the one based on values of the currently identified subsystem. The adaptation is ended if some other subsystem needs a criterion. A simple criterion often used is the absolute value or square of the control error

$$J > eps \quad J = \max_r |e_r(k)|; \quad r \neq i \tag{8.5}$$

where eps is a threshold for the decision whether control of the r-th is accurate enough or not, e_r is the control error of the r-th subsystem and k is the current identification step. The criterion is evaluated for each subsystem except the currently identified one. The index corresponding to the currently identified subsystem is i. Exceeding the criterion could be a sign that the controller is not properly tuned and identification of the corresponding subsystem should be performed.

Indicators that are more complex can be used for the criterion instead of the control error, *e.g.* an integral criterion of the quality of the control process.

Observing the outputs or control errors of other subsystems corresponds better to the requirement of industrial applications of decentralized control. It is usually not necessary to reach the exact reference signal value but it is required, that all outputs reach predefined tolerances. The disadvantage of this approach is that it does not contain any effort to avoid problem states – adaptation is not ended as long as no outputs exceed preset limits.

Choice of Next Controller for Adaptive Mode

After adaptation of a particular subsystem is ended, the supervisory system has to solve the task of selecting the next subsystem to switch to adaptive mode *i.e.* to move the identification token to. Often a preset sequential subsystem order is set: after adaptation of the first subsystem is ended, the token is moved to the second subsystem, then to the third subsystem and so on. After ending adaptation of the last subsystem, the token is moved back to the first subsystem.

Better results can be achieved if the next subsystem to be switched to adaptive mode is determined dynamically during the control process. A typical case when this approach is used is the case of ending adaptation on the basis of the parameters of other subsystems. In this case, the token is naturally moved to the subsystem which caused the ending of adaptation.

8.1.3 Logical Supervisor

A *logical supervisor* has been proposed to utilize and simplify the design of a supervisory system. This approach is suitable for use in real-time industrial applications. The idea of a logical supervisor is based on the following two principles:

- assigning priorities to individual subsystems;
- on-line evaluation of criteria for each subsystem.

The situation that reaching the reference value is more important for some subsystems than for others is very common, especially in industrial applications. It is thus possible to assign a unique priority to each subsystem. The priority corresponds to the importance of the subsystem's output. The numbering of subsystems is just a formal problem and thus the subsystems can be numbered according to priorities. The first subsystem has the highest priority; the second subsystem has the second highest priority and so on.

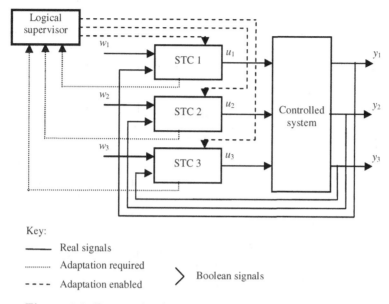

Key:

──── Real signals

·········· Adaptation required

- - - - Adaptation enabled   ⟩ Boolean signals

Figure 8.2. Decentralized control circuit with logical supervisor

Further, for each subsystem, a criterion which determines whether the subsystem requires switching to adaptive mode or not, is calculated. The criterion can be designed with respect to particular properties of the subsystem. The block responsible for computing the criterion can be encapsulated with the self-tuning controllers and the output, which is sent to the logical supervisor, is a Boolean value determining whether or not the subsystem requires adaptation.

The last part of the logical supervisor approach is a superior logic determining which of subsystems requiring adaptation will be switched to adaptive mode. The decision-making is based on priorities assigned to individual subsystems. If the first subsystem requires switching to adaptive mode it is always satisfied; if the second subsystem requires switching to adaptive mode, it is satisfied only if the first subsystem does not require switching to adaptive mode, *etc.* The control circuit schema with logical supervisor approach and a controlled system of three inputs and three outputs is shown in Figure 8.2 when only one of the dashed signals, "Adaptation enabled", is switched on at a time.

The logical supervisor uses only the logical values on its input and provides logical values on its output. In addition, the relations between inputs and outputs are simple logical functions. The transfer function between the input and output signals of the logical supervisor can be arranged as a table of logical values. This situation is shown in Table 8.1 for the MIMO system of three inputs and three outputs.

Table 8.1. Relations between inputs and outputs of a logical supervisor

Required adaptation (inputs)			Enabled adaptation (outputs)		
R_1	R_2	R_3	E_1	E_2	E_3
0	0	0	0	0	0
0	0	1	0	0	1
0	1	0	0	1	0
0	1	1	0	1	0
1	0	0	1	0	0
1	0	1	1	0	0
1	1	0	1	0	0
1	1	1	1	0	0

It is also possible to rewrite relations between inputs (R_k) and outputs (E_k) using logical operators:

$$E_1 = R_1$$
$$E_2 = \bar{R}_1 \text{ AND } R_2 \tag{8.6}$$
$$E_3 = \bar{R}_1 \text{ AND } \bar{R}_2 \text{ AND } R_3$$

where the bar denotes negation of a variable and function AND represents logic product. When determining signal "Adaptation enabled" for a general controlled MIMO system, the following relation is valid:

$$E_k = \prod_{i=1}^{k-1} \bar{R}_i \text{ AND } R_k \tag{8.7}$$

where $\prod$ stands for logic product.

The logical supervisor represents a reliable approach to the design of a supervisory system for decentralized control. The advantage of this approach

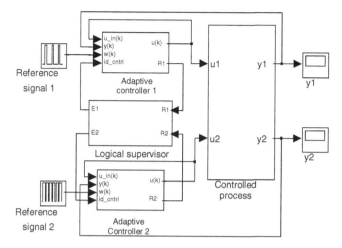

Figure 8.3. Simulink control circuit with TITO controlled system

is its simplicity of implementation and the small number of signals that are transferred from subsystems to supervisory system and back.

The logical supervisor was tested in connection with controllers from the STCSL. The properties of controllers were tested in the MATLAB® or SIMULINK® environment using the control scheme shown in Figure 8.3. The model contains a continuous-time TITO system controlled by two self-tuning controllers and a logical supervisor which provides identification switching between input/output pairs.

The quality of the control process is affected by many parameters; *e.g.* sampling period, control law algorithm, logical supervisor algorithm, saturation, initial parameter estimates. The recursive identification uses a least squares method with adaptive directional forgetting.

8.1.4 Control of Air Heating System Model

The air heating system model is shown in Figure 8.4. This system has two inputs (rotations of ventilator and power of resistance heating) and two outputs (the flow of air, measured by the rotations of an airscrew; and temperature inside the tunnel, measured by a resistance thermometer). The system was divided into two input/output pairs, the first pair consisting of ventilator rotations as input u_1 and flow of air as output y_1. The second pair consist of resistance heating power as input u_2 and temperature as output y_2. The aim is to control a TITO system using two standalone single-input/single-output controllers; *i.e.* a decentralized approach was applied.

The connection of the laboratory model and the SIMULINK® environment has been realized through control and measurement PC card Advantech PCI-1711. Blocks for reading analogue inputs and for writing to the analogue

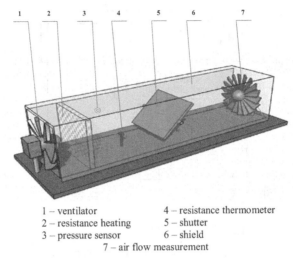

1 – ventilator   4 – resistance thermometer
2 – resistance heating   5 – shutter
3 – pressure sensor   6 – shield
                    7 – air flow measurement

Figure 8.4. Laboratory air heating model

outputs on the PC card were used to communicate with the model. These blocks are implemented as s-functions written in C language, which allows low-level access to the ports of the PC computer. This mechanism allows the connection of SIMULINK® and any PC compatible equipment designed to collect external data.

The control performances using pole assignment controllers *pp2b_1* (4.135) given by

$$u(k) = - [(q_0' + \beta)y(k) - (q_0' + q_2')y(k-1) + q_2'y(k-2)]$$
$$-(\gamma - 1)u(k-1) + \gamma u(k-2) + \beta w(k) \tag{8.8}$$

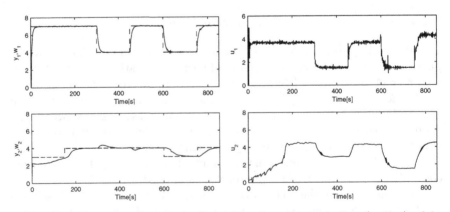

Figure 8.5. Control performance using pole assignment controllers (*pp2b_1*) – laboratory heating model

are shown in Figure 8.5. The controller output variables are the air flow source (speed of the ventilator – u_1) and the heat source (resistance heating – u_2). The process output variables are the speed value of the air indicator – y_1 and the air temperature – y_2. The following controller parameters were chosen:

First controller: sampling period $T_0 = 1$ s, damping factor $\xi = 3$, natural frequency $\omega_n = 1$.

Second controller: sampling period $T_0 = 1$ s, damping factor $\xi = 10$, natural frequency $\omega_n = 1$.

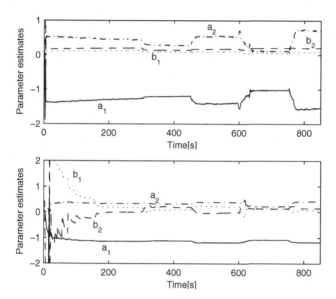

Figure 8.6. Evolution of parameter estimates using *pp2b_1* controllers – laboratory heating model: upper graph – control of air-flow, lower graph – control of temperature

The time constant of the pair ventilator–flow is relatively small compared with time constant of the heating–temperature pair and thus the output of the first pair becomes steady earlier. Results demonstrate that the change of resistance heating power does not affect the flow of air, but the influence of ventilator rotation on the temperature is substantial. The parameters of the heating–temperature controlled system thus change over time and the on-line identification method used should assign greater weight to newer data than to older data. The evolution of parameter estimates are presented in Figure 8.6. It is obvious from Figure 8.6 that changes of parameter estimates are greater in the first phase, caused by inaccurate initial parameter estimates. Further changes of parameter estimates correspond to changes of reference signal, indicating the presence of a nonlinearity in the system.

The control performance was different when a controller of another type was used, as shown in Figure 8.7. The evolution of the reference signal is the same as in the previous case, but *zn2br* controllers (4.16) given by

$$u(k) = K_P \left\{ e(k) - e(k-1) + \frac{T_0}{T_I} e(k) \right.$$
$$\left. + \frac{T_D}{T_0} [e(k) - 2e(k-1) + e(k-2)] \right\} + u(k-1) \tag{8.9}$$

were used. The *zn2br* controller uses the Ziegler–Nichols algorithm to compute the value of controller output using a backward rectangular method of discretization integration component with sampling period $T_0 = 1$ s.

In this case process output oscillations are significantly greater and the quality of control is lower compared with the pole assignment controllers.

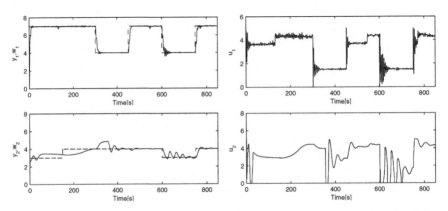

Figure 8.7. Control performance using Ziegler–Nichols based controllers (*zn2br*) – laboratory heating model

In all cases the initial vector of parameter estimates was chosen without *a priori* information $\hat{\Theta}^T(0) = [0.1, 0.2, 0.3, 0.4]$ and controllers outputs were limited within the range $\langle 0; 5 \rangle$.

8.1.5 Control of Coupled Motors CE 108

Another practical verification of decentralized control using STCSL in connection with the Real-Time Workshop was carried out with coupled servomotors CE 108 (producer TecQuipment Ltd., Nottingham, UK). The schema of this model is shown in Figure 3.8 (see Section 3.2.4).

This system has two inputs (rotations of the left, u_1, and the right, u_2, servomotors) and two outputs: the speed, y_1, of the belt measured by pulley rotation and the tension, y_2, of the belt measured by deviation of the jib. Considering the mechanical point of view, the system was divided into two

input/output pairs, the first pair consisting of the rotations of the left servomotor, input u_1, and the tension of the belt, output y_2. The second pair
consists of the rotations of the right servomotor, input u_2, and the speed of the
belt, output y_1. This system is strongly nonlinear with great interactions between individual loops. The static characteristics of this equipment are shown
in Figure 3.10.

A sample of the control performance using Ziegler–Nichols based controllers *zn2fr* with sampling period $T_0 = 0.25$ s given by Equation (4.18)

$$u(k) = K_P \left\{ e(k) - e(k-1) + \frac{T_0}{T_I} e(k-1) \right.$$
$$\left. + \frac{T_D}{T_0} [e(k) - 2e(k-1) + e(k-2)] \right\} + u(k-1) \tag{8.10}$$

is shown in Figure 8.8. The first part of the control sequence (approximately
0–40 s) is used to adapt controllers to the system because the initial model
parameter estimates are set without *a priori* information about the controlled
system.

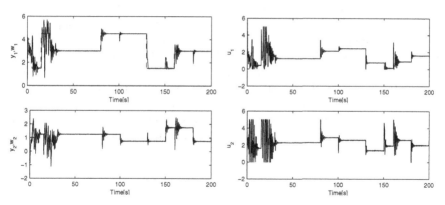

Figure 8.8. Control performance using Ziegler–Nichols based controllers (*zn2fr*) –
coupled motors CE 108

Another control performance is shown in Figure 8.9. The pole assignment
controllers *pp2a1* given by equation (4.120)

$$u(k) = q_0 e(k) + q_1 e(k-1) + q_2 e(k-2) + (1-\gamma)u(k-1) + \gamma u(k-2) \tag{8.11}$$

were used in this case. The poles were set so that the closed loop behaves like
a second-order continuous-time system with controller parameters in both
circuits $T_0 = 0.1$ s, $\xi = 1$ and $\omega_n = 2$.

In all cases the initial vector of parameter estimates was chosen without
a priori information $\hat{\Theta}^T(0) = [0.1, 0.2, 0.3, 0.4]$ and controller outputs were
limited within the range $\langle 0; 5 \rangle$.

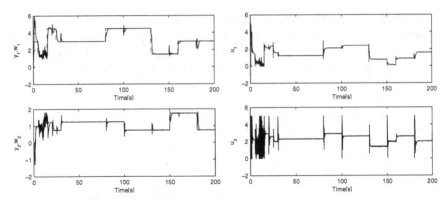

Figure 8.9. Control performance using pole assignment controllers (*pp2a1*) – coupled motors CE 108

8.1.6 Control of Twin Rotor MIMO System – Helicopter

The final laboratory model used to verify the decentralized approach to the control of multivariable systems, is the Twin Rotor MIMO System (Feedback Instruments, Ltd, UK). This model, which is shown in Figure 8.10, provides a high-order, nonlinear system with significant cross-coupling. The main parts of the system are the pedestal, the jib connected to the pedestal and two propellers at the ends of the jib. The system jib can freely rotate around the vertical axis by about 330 degrees (process output $y_1(t)$) and the horizontal axis by about 100 degrees (process output $y_2(t)$). The system inputs $u_1(t), u_2(t)$ are the voltages used to drive the propeller motors, and outputs are angular rotations with respect to horizontal and vertical axes.

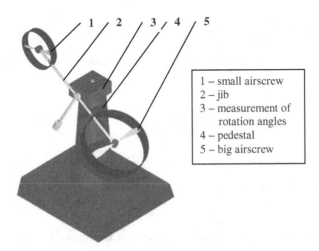

Figure 8.10. Twin rotor MIMO model (helicopter)

Despite the strong interactions in the system, decomposition to subsystems is straightforward:

- the first subsystem consists of the small propeller which drives the angular rotation around the vertical axis;
- the second subsystem consists of the big propeller driving the angular rotation around the horizontal axis.

Before the control circuit was connected as a closed loop, experiments to obtain the static characteristics of the systems were performed. The influence of the first system input on the second system output is small but the evolution of the second output is a sign of nonlinearity of the system. Another control problem using this system is the large hysteresis which is present in the system. The static characteristics of the first subsystem, measured for increasing and decreasing input signals, are shown in Figure 8.11. The great influence of changes of the second system input on the first output was confirmed by this measurement.

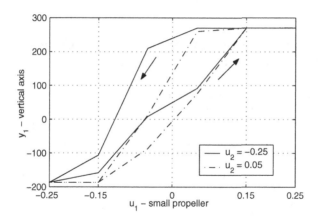

Figure 8.11. Static characteristics of the first subsystem showing hysteresis – twin rotor MIMO model

The control performance of the twin rotor MIMO system using *pp2b_1* controllers (8.8) is presented in Figure 8.12. The following controller parameters were chosen:

First controller: sampling period $T_0 = 0.5$ s, damping factor $\xi = 10$, natural frequency $\omega_n = 1$.

Second controller: sampling period $T_0 = 1$ s, damping factor $\xi = 10$, natural frequency $\omega_n = 1$.

In both cases the initial vector of parameter estimates was chosen without *a priori* information $\hat{\Theta}^T(0) = [0.1, 0.2, 0.3, 0.4]$. Figure 8.12 demonstrates that this strongly nonlinear, unstable and high-order system can be stabilized, and

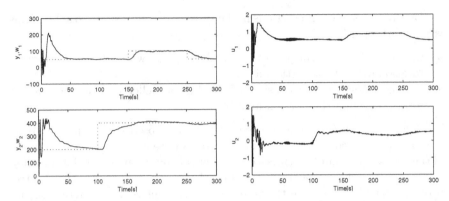

Figure 8.12. Control performance using pole assignment controllers (*pp2b_1*) – twin rotor MIMO model

also that quite good asymptotic tracking can be achieved by using adaptive control without *a priori* information about the process model.

8.2 Application of the Adaptive Linear Quadratic Controller to a Heat Exchanger Station

A team from the Faculty of Mechanical Engineering of the Slovak University of Technology in Bratislava prepared the experimentation on the heat exchanging station at the university. A PC computer was connected to the existing control system. Automatic switching mechanism was build up to switch on the standard controller in case the experiment leads to the undesirable behaviour. In this way the problems with the technology as well as operators of the plant were safely prevented. The MATLAB® SIMULINK® environment was used for the communication with the technology. Various predictive controllers were tested in the plant and are reported in [133]. Here an experiment with a LQ adaptive controller is described.

8.2.1 The Technology

Trials were carried out on a standard industrial heat exchange station at the university campus. The station supplies heat and hot water to all university buildings and facilities. It is equipped with a two-stage tubular concentric counterflow exchanger having 20m² of heat-exchanging surface. The temperature of the outlet is controlled by varying the flow of the primary hot water by means of a servovalve. The dynamics of the exchanger can be roughly approximated by a first-order system with time delay about 30 s and rise time about 120 s. In the existing setup, the plant is controlled by a PID controller running at a sampling period of 0.5 s and with an outlet temperature set point of 55°C. The PID constants are as follows: $P = 2, I = 0.003, D = 0$.

8.2.2 Linear Quadratic Controller

The adaptive LQ controller used in the experiment was described in Chapter 6, and the LQ toolbox in Section 7.2. Its setup is as follows:

1. The bsic model of the process was assumed to be in the form of second-order regression model

$$y(k) = -\sum_{i=1}^{n} a_i y(k-i) + \sum_{i=0}^{n} b_i u(k-i) + e_k + \sum_{i=1}^{n} d_i v(k-i) + K \quad (8.12)$$

but other structures were tested as well, typically first- and third-order and less time delay.

2. The control quality was evaluated by the criterion

$$\Psi = \sum_{i=k+1}^{k+T} [f_y(w(i) - y(i))^2 + f_u(u(i) - u0(i))^2]. \quad (8.13)$$

This penalizes the control error between the actual and desired output and the difference between the actual input and some reference $u0$. The introduction of this reference is a novelty and was discussed in Chapter 6; f_y and f_u denote for the data-dependent penalty. This form of criterion widens substantially the possibilities of controller tuning and reflects the practical situation when quite often a function of the input or output is required to fulfil a given specification.

3. The minimization of criterion (8.13) based on model (8.12) is based on dynamic programming in state space. The nonminimum dimension state space is used here, using delayed measured data as elements of the state space. This choice increases only moderately the computational burden of the square root minimization algorithm, simplifying the interpretation of results considerably.

4. A reliable square root algorithm for minimization of the criterion (see Section 6.7) allows simple implementation of the various types of penalization considered in the criterion.

5. The IST (iteration spread in time) (see Section 6.3.2) strategy ensures the adaptive version of the approach is sufficiently fast.

6. To be able to cope with typical signal constraints the controller algorithm enables automatic penalization adjustment based on constraints of the input signal and its moves [119].

7. The introduction of input reference signal $u0$ is used to smooth the transition from the existing PID controller to the adaptive LQ controller.

8. A standard on-line LS algorithm with exponential forgetting was used for system parameter identification.

8.2.3 Programming Aspects

All the features listed above have been implemented into MATLAB® and SIMULINK® functions. Basic SIMULINK® schemes used for the direct control of the plant are shown in Figure 8.1 and Figure 8.2. The first one realizes the positional control, the second one adds an integrator to the loop.

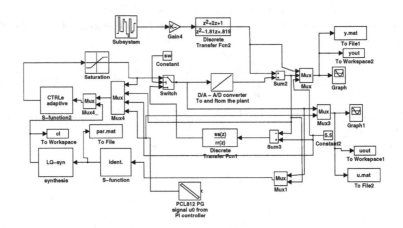

Figure 8.13. Control of the plant with incremental LQ adaptive controller

Two types of controller with special properties and automatic adjustment of the criterion were prepared and tested. First, was a controller which operates only in cases when the disturbance exceeds a given level. Such a controller can be called "cheap" as it does very little, more precisely, it keeps the input at a level $u0$. A scheme representing a "cheap" controller in which penalization of the output depends on the predicted control error is schown in Figure 8.14. The output of the block is system input $u(t)$, the input points in_1 to in_6 are places where the following signals enter

in_1 – system input $u(t)$,
in_2 – system output $y(t)$,
in_3 – set point $w(t)$,
in_4 – output level where the output starts to be penalized,
in_5 – minimum penalization (when no additional penalization is used),
in_6 – the pair of $[u(t), y(t)]$ for parameter estimation.

Another example of a controller with data-dependent penalizations, in which the input penalization depends on the level of the input itself and in such a way that it suppress large input values, was prepared. Figure 8.15 shows the block diagram of the arrangement for data-dependent input penalization adjustment. Here the inputs points of the block are:

in_1 – system input $u(t)$,
in_2 – system output $y(t)$,
in_3 – setpoint $w(t)$,
in_4 – reference input $u0(t)$,
in_5 – minimum penalization (when no additional penalization is used)
in_6 – the pair of $[u(t), y(t)]$ for parameter estimation.

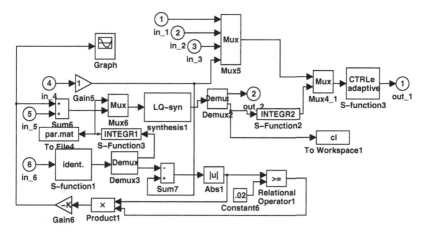

Figure 8.14. Adaptive LQ controller with output penalization depending on the output prediction – "cheap" controller

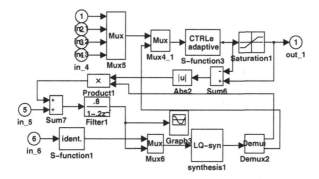

Figure 8.15. Adaptive LQ controller with input penalization depending on constraints

As the setting of penalizations are in both cases totally independent they can be simply implemented together.

Figure 8.16 shows the way of using the reference input signal. The signal can be either set to some value, or it can be linked to the output of the original controller.

8.2.4 Experimental Results

Experiments were carried out in three periods of the year representing different operating conditions of the technology.

The aim was to test:

- behaviour of the LQ controller, *i.e.*

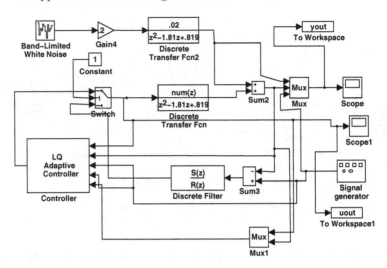

Figure 8.16. The use of adaptive and fixed default controllers together

- how the input penalization influences the behaviour
- the influence of the structure of the model (controller)
- the possibilities of more sophisticated penalizations;
- adaptation process;
- startup of the adaptation;
- ways to ensure smoothness of the transition from standard control to LQ and back.

The first experiments were carried out in the summer period. The primary water temperature was low (70°C) and hot water consumption was rare. To see the controller action artificial disturbances were created by opening a hot water tap for approximately one minute. Several experiments were run on the heat exchanger: different models, sampling and penalization were tested in these cases. Using prior information about the plant a second-order regression model with two time delay steps for the sampling period $T_0 = 10$ s was used initially. A few experiments showed that this model was satisfactory, and that a reasonable penalization was within the interval $Q_u = (0.1 - 1)$. In all the following figures the dotted lines indicate reference values, *i.e.* the set point for the output and the reference input for the input. In most cases this reference input is an output of the standard controller.

Figure 8.17 shows the situation when the adaptive controller starts in a smooth way from the standard PID controller of the plant. The dropouts in the input variable were caused by stopping the simulation to change penalization Q_u. Then the simulation started with zero initial conditions. The output variations were caused by a disturbance. This can be deduced from the fact that both the PID and adaptive controllers tend to change input value at

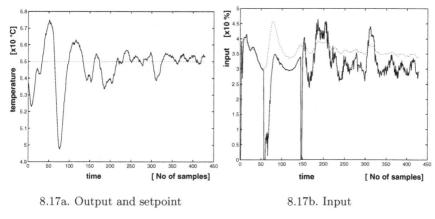

8.17a. Output and setpoint   8.17b. Input

Figure 8.17. Experiment No 1

the same direction. When dropouts occurred the PID controller moved in the opposite direction.

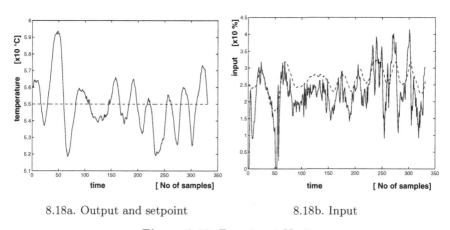

8.18a. Output and setpoint   8.18b. Input

Figure 8.18. Experiment No 8

Figure 8.18 documents again the nice startup given by smooth tuning from the standard PID controller to the full LQ controller.

Table 8.2 shows the identified roots of the numerator and denominator in various experiments. The second part of the table contains the results from the second period (only one part of a complex pair is shown).

More interesting results were obtained from the third, winter, period. Experiment No 13 used a shorter sampling period. The standard structure was

Table 8.2. Roots α_i of denominator and β_i of numerator of identified model

Exp. No	α_1	α_2	β_1	β_2	β_3	β_4
1	0.9953	0.3163	0.60-0.53i	0.60-0.53i		
2	1.1004		1.17	0.12+0.79i	-0.64	
4	0.9960	0.4809	-6.25	1.40	0.57+1.10i	
7	0.9968	0.2723	-8.58	1.11	-0.57+0.38i	
8	0.9854	0.7607	1.17+0.52i	0.57+0.75i	0.41+0.80i	-0.46
9	1.0226	0.7924	1.81	1.40	37+1.04i	-0.45+0.69i
10	0.9821	0.7651	1.94+0.49i	-0.11+0.36i		

used but with sampling $T_0 = 5\,\text{s}$. The input and output are shown in Figure 8.19, and the identified roots of the numerator and denominator in Figure 8.20.

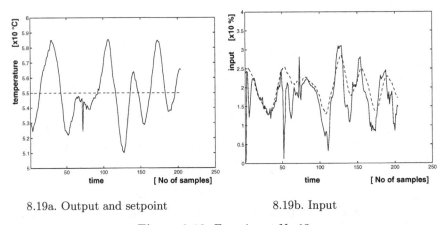

8.19a. Output and setpoint   8.19b. Input

Figure 8.19. Experiment No 13

Experiment No 14 lasted more than 2 hours and contained the lunch time rush hour. The temperature control is very good (Figure 8.21).

Figure 8.22 documents a very quiet startup from zero initial conditions.

Experiment No 18 was carried out with the scheme shown in Figure 8.14. The details of the input and corresponding data-dependent penalization are shown in Figure 8.24.

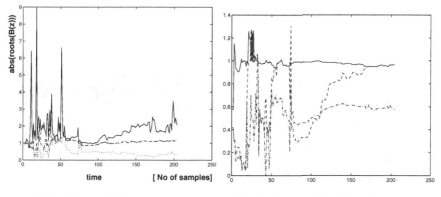

8.20a. Roots of the numerator   8.20b. Roots of the denominator

Figure 8.20. Experiment No 13.

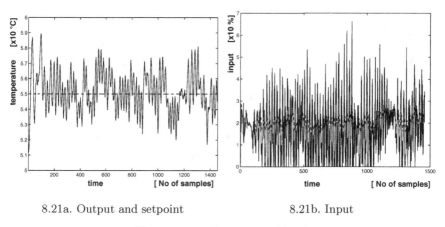

8.21a. Output and setpoint   8.21b. Input

Figure 8.21. Experiment No 14

8.2.5 Conclusions

Experiments with LQ controllers in the temperature loop of a heat exchanger were performed in three periods with different loads and disturbances. In all cases the adaptive LQ control was successful and the input penalization Qu was sufficient to tune the behaviour of the controller from "smooth", when the quality is similar to a PID controller, to "restless", however, achieving better quality. The identification results confirmed the second-order model structure with time delay. The choice of time delay was not crucial. The examples with rare disturbances show how the parameters tend to a first-order structure and disturbance moves them back to second-order. A difference in denominator of the model transfer function obtained in different periods can be seen. In all cases the resulting model was nonminimum phase, which

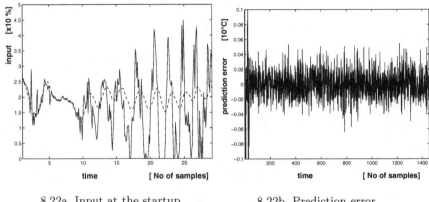

8.22a. Input at the startup   8.22b. Prediction error

Figure 8.22. Experiment No 14

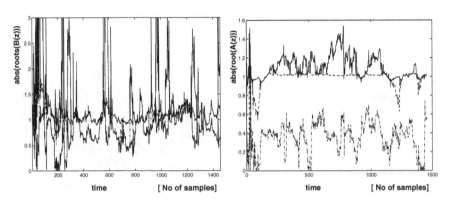

8.23a. Roots of the numerator   8.23b. Roots of the denominator

Figure 8.23. Experiment No 14

explains the discrepancy between prediction errors of the model shown and
the resulting controller behaviour. The experiments have shown that for a
given purpose a simple PID (in fact, practically, only P) controller is fully
satisfactory for the desired quality. On the other hand, these experiments
verified that the application of an adaptive controller could also be simple
and safe. The quality of control obtained was not much better due to existing
delays and nonminimum phase behaviour, but it could be expected that if the
disturbances could be measured, the control performance could be improved.

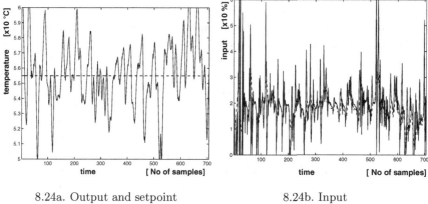

8.24a. Output and setpoint   8.24b. Input

Figure 8.24. Experiment No 18

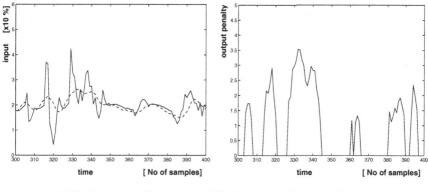

8.25a. Input detail   8.25b. Corresponding penalization of output

Figure 8.25. Experiment No 18

8.3 Boiler Control With Multivariable Linear Quadratic Controllers

Control loops of the outlet steam pressure and outlet superheated steam temperature are usually the most important loops in steam boilers. Their dynamics vary with steam flow (load) and working conditions and depend also on acting disturbances. Adaptive LQ controllers were applied to control simultaneously outlet steam pressure and two outlet superheated steam temperatures.

8.3.1 The Technology

In the power plant of a metallurgical factory, two fossil-fuel-fired drum boilers (with the same parameters: 120 ton/h, 9.5 MPa = 95 bar, 540°C) deliver steam into a common collector. The steam is conducted into a 25 MW back-pressure turbine and to various metallurgical shops to be used for production purposes. The steam boiler has two branches of superheaters. The steam temperatures behind the last superheaters of both branches are controlled by injected water flow through spray valves before the superheaters. The steam temperature is influenced by steam pressure, variable load, and fuel changes and by the injected water in the valves. This suggests strong interaction between both branches. Changing the speed of powdered coal feeders controls steam pressure. According to the load, the number of coal feeders used varies from 4 to 8. The powdered coal intake is measured only indirectly by the positions of actuators on the speed changing units and depends also on the state of the coal bunkers. The quality of coal and its calorific value varies too. Therefore the dependence of coal intake on feeder speed is time variable. The combustion air intake is controlled by turning the control rims of two fans. Adaptive control of the combustion airflow is desirable because the coal intake and quality varies for the same load (steam flow). The common steam collector means that control of the steam pressure in one boiler is strictly influenced by conditions in the other boiler, by the momentary condition of the combustion process, and by changes of load. The normal load of each boiler is about 100–110 ton/h.

An adaptive LQ controller (LQ Self-tuning Controller, LQ STC) was experimentally applied as a multi-variable (MIMO, multi-input, multi-output) STC to control simultaneously the following loops of the drum boiler plant [129]:

(A) outlet steam temperature on the left branch of the final superheater;

(B) outlet steam temperature on the right branch of the final superheater;

(C) outlet steam pressure with an additional adaptive control loop for the combustion air (a special algorithm was designed for combustion air control).

8.3.2 Programming Aspects

The LQ STC was programmed as a module composed of several subroutines called by up to four independent programs. Therefore it was possible to use simultaneously up to four LQ STC in a MIMO version. Such a solution enabled the use of one or more LQ STC programs to check proper model structure, control synthesis *etc.* and others directly for control. A multivariable regression model of Nth-order with n process inputs $u(k)$ and n process outputs $yc(k)$ and with r measured external disturbances $v(k)$ was programmed; these external disturbances were part of an $(n+r)$-dimensional extended vector output $y(k)$. All measured values were related to reference values. Constant values (equal set point) were used as reference values for process outputs. Constant values or the preceding measured values could be taken as reference values for inputs, that is "incremental" inputs were considered as increments of measured values. Zero initial parameter values were used except the value of parameter $B0$.

Exponential forgetting was applied for identification. The sum of squares of the prediction error was calculated and used in model comparison.

Controller algorithm

Estimated parameters $\Theta(k)$ or also their covariance matrix $C(k)$, *i.e.* uncertainty of parameters (in so-called *cautious control*) were used for control synthesis. The use of uncertainty of parameters $C(k)$ leads to an additional control penalisation varying according to $C(k)$. A *mMoving horizon strategy MH* was used with only final state penalisation, as well as *an iteration spread in time strategy IST*; see Section 6.3.2. The horizon (number of Riccati equation iterations) was chosen according to convergence of the control law

$$|Cl(i+1) - Cl(i)| < eps * |Cl(i+1) + Cl(i)|$$

where *eps* is a chosen constant. Additionally, maximum number of iterations (steps) M was allowed.

Calculated controller output $u(k)$ was limited; the upper and lower limits varied (with an exponential rate, from initial values equal to the reference value at start time) up to their maximum values. In this way an initial bump at the start of STC was efficiently excluded.

8.3.3 Connection of the Adaptive Controller into the Existing Control Loop

The adaptive controller calculated the manipulated variables $u(k)$ as the desired value (set point) for the inner loops of the actuators. These actuators were controlled in a servo-loop by digital P regulators using pulse width modulation in the case of computer control. The sampling periods of both inner and outer loops were the same, 20 s. The vector of inputs $u(k)$ to the system (*i.e.* manipulated variables) consists of:

- the temperature behind the spray valve for steam temperature control;
- the position of one coal feeder actuator for steam pressure control.

The remaining feeders were set to the same desired value. The desired values for both air fans were calculated according to the desired value of coal feeders with the adaptively varying relation between air and coal (this relation was based on the estimated parameters of the steam pressure model applied by LQ STC).

Note: Input Definition and Connection
As the input is usually provided by an actuator in the inner loop, it is possible to choose the system model input $u(k)$ as

(A) the desired value of manipulated variables calculated by LQ STC (set point for the inner loop with the actuator), where the inner loop dynamic including actuator becomes part of the whole identified system; or

(B) the measured value of manipulated variables (controlled variable of the inner loop), where only 'own system' is identified on the basis of the true acting manipulated variable, and the inner loop dynamic is not considered part of the identified system.

In this application only case (B) was used. The pros and cons of the two variants are discussed in [64].

8.3.4 Control Results

The existing analogue control of the control loops did not work well under changing process conditions and therefore these loops were usually controlled manually on both boilers. The failures of coal intake, the interaction of both boilers and rapid changes in outlet steam flow generallyled to load changes on the boiler of about 20 ton/h, but changes up to 50 ton/h also occurred.

As there was no previous knowledge of the controlled system and no previous identification results then all "structural parameters" of the LQ STC (model order N, the choice of external measured disturbances, control strategy, the type of variable – increments or not, exponential forgetting value, $etc.$) had to be determined directly on the plant. The predicted error performance was used as the criterion for model structure comparisons and the control error quadratic criterion for the control strategy. Models with various structural parameters were identified at the same time and compared under the conditions of running LQ STC. No artificial signals were used, only the real control action of LQ STC. (Four LQ STC programs could be used simultaneously.) Results of such tests led to the use of adaptive controllers LQ STC always realized with measured disturbance $v(k)$ and with moving horizon strategy only.

The steam temperatures of both branches were controlled alternatively by

- two independent single-input single-output (SISO) controllers with measurable external disturbance included in the extended output $y(k)$,
- one multi-input multi-output (MIMO) controller with two-dimensional input and controlled output $yc(k)$ and with two measurable external disturbances included in the extended output.

Steam pressure or steam flow and air fan position were used as measurable external disturbances. The control results with the MIMO temperature controller were better than with the two SISO controllers. The temperature behind the superheaters was controlled over a smaller range with less control effort. The MIMO controller with regression model of order $N = 3$ and two measurable external disturbances led to the estimation of 80 parameters. Inputs $u(k)$ use constant reference values. A moving horizon strategy was used and the uncertainty of parameters incorporated. A cautious control strategy led to better results than the strategy based on certainty equivalence (smaller range of control errors and overshoots). Therefore the temperature set points were increased.

The steam pressure was controlled by a SISO controller together with combustion air intake. The model order was $N = 3$ with two measured disturbances (steam flow and air fan position): 39 parameters were estimated. Again the moving horizon strategy was used. The use of a model with incremental inputs improved steam pressure control performance. The STC had higher and faster adaptability to various nonlinear and nonstationary process changes including fast load changes and a jump change in the number of feeders in operation. The significantly lower range of control errors and smaller overshoots in the case of larger process and load changes enabled the steam pressure set point to be increased from the usual 93 bars to 95 bars. Improved boiler control, including improved combustion control, favourably affected the operation of both boilers and the turbine power. This led plant operators to use adaptive controllers as standard controllers.

The convergence of the Riccati matrix was fast. The number of steps $M = 5$ for steam pressure control or $M = 3$ for MIMO temperature control provided sufficient horizon for the criterion. The sampling period $T_0 = 20$ s was used in all control loops.

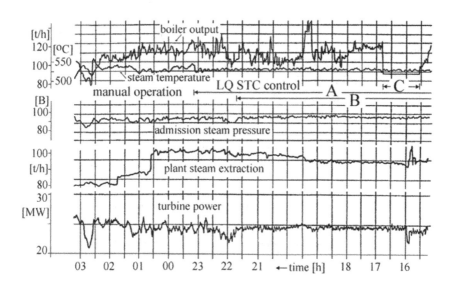

Figure 8.26. Drum boiler control with LQ STC and manual operations. (A) Two-variable control of steam temperature loops with MIMO LQ STC. (B) steam pressure control including combustion optimization with SIMO LQ STC. (C) Failure in the measurement of steam flow, *i.e.* boiler output (external variable of LQ STC)

Typical control responses with and without a LQ STC to control all three loops by two LQ STCs are documented by plant plotter records shown in Figures 8.26 and 8.27. The boiler steam flow (load) and the steam temperatures

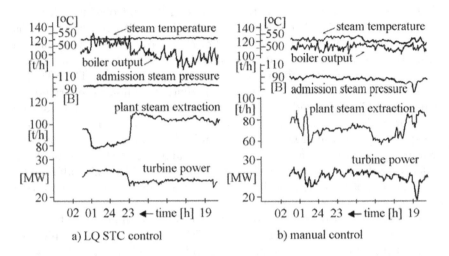

a) LQ STC control   b) manual control

Figure 8.27. Drum boiler control: LQ STC and manual operations

are recorded, together with variables measured on the turbine: the amount of steam led into the plant (labelled steam extraction), the turbine power and admission steam pressure. The second boiler delivers steam into the same common collector and therefore admission steam pressure shows pressure of steam blended from both boilers. Start of LQ STC from zero initial parameter values (except parameter $B0$) was bumpless (see Figure 8.28); no troubles occurred at the start due to applied variable output limits and relatively large forgetting of data (value $frg = 0.97$ to 0.98) enabling fast identification of model parameters from their initial zero values. LQ STC also started directly from last parameter values, mainly when turned on by plant operators.

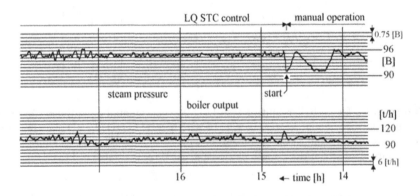

Figure 8.28. Start of LQ STC. Start from zero initial parameter values except parameter B0

A failure in the measurement of one of the measured disturbances (boiler steam flow), which occurred several times, produced no significant deterioration of steam pressure control. Covariance wind-up did not take place, probably due to the nonstationarity of process model parameters, the applied multistep control criterion, and permanent process changes. Later the applied regression models were compared with ARMAX models on an external computer. Identification from data files obtained under a standard control strategy led to almost the same values of performance criterion for both types of model.

This application was one of the first MIMO STC applications and helped with other later applications.

8.4 Adaptive Linear Quadratic Controller in Cascade Control

Control of superheated steam temperature in a cascade control loop using an adaptive LQ controller in the power plant was described in [134].

8.4.1 The Technology and Controlled Loops

Long-term experiments were done on a fossil-fuel-fired once-through boiler of 200 MW (660 t/h, 540°C) power plant to control superheated steam temperatures. This boiler has superheated system divided into two branches (A and B), each branch consisting of three superheaters (PI, PII and PIII). Outlet superheated steam temperatures from both final superheaters PIII (labelled T20A and T20B) are mixed and fed into a turbine. The steam temperature behind each superheater is controlled in a cascade control loop. The superheater is a system directly influenced by changes of inlet steam flow and temperature, sprayed water and heat flow. The approximate time constant of such a system varies significantly with load. Therefore conventional PI controllers are not always suitable to control steam temperature under varying conditions and in the presence of large process changes and failures which result in large temperature overshoots. The adaptive control was verified on the right branch "B" only. This branch "B" had worse dynamics than branch "A". The left branch "A" remained controlled via standard PID analogue control to compare control results [134]. Two different control concepts were applied. In the first, the main analogue PI controller of the cascade control loop was replaced by the LQ STC. In the second concept the LQ STC was connected and added into the control of the two final superheaters and calculated a set point for the main analogue PI controller. The following applications of the LQ STC were proved:

(A) *The control of steam temperature behind the superheater PII*
Superheated steam temperature behind the superheater is controlled in the

following manner (Figure 8.29). The main (master) analogue PI controller
R1 of the cascade control loop calculates set point T16A, T16B, of the
steam temperature behind the spray valve (*i.e.* the steam temperature
before the superheater T16AZ, T16BZ) for the small (slave) loop with
analogue PI controller (R2). The controller R2 sets a signal for the servo
loop of the actuator which operates the spray valve. The main controller R1
in the cascade control loop of the second superheater (PII) was replaced
by the LQ STC. The small loop and servo loop remained controlled by
analogue controllers. The set point of the main controller R1 (T18AZ,
T18BZ) is constant as usual.

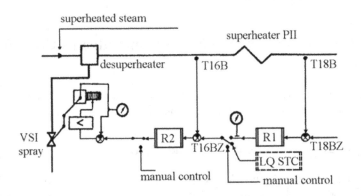

Figure 8.29. Diagram of superheated steam temperature control behind the superheater PII

(B) *The control of steam temperature of the final two superheaters*
    The STC was connected into the control loop of two final superheaters
    (PII and PIII). The LQ STC calculated the set point T18BZ for the second
    superheater PII on the basis of the control error of the final superheater
    (PIII). The whole analogue cascade control for PII and PIII remained in
    operation (Figure 8.30). However, now the superheater PII was controlled
    to the variable set point T18BZ to keep the controlled variable, *i.e.* the
    steam temperature T20B behind the PIII.

8.4.2 Programming Aspects

LQ STC was programmed as a MIMO controller with measured external
disturbances similarly to the previous application. The controller algorithm
incorporated new theoretical results. Not only final state penalization but
direct input penalization or input differences (increments) penalization were
used. Conditional interruption of identification was applied in the case of small
or very large prediction errors $e_p(k)$. If

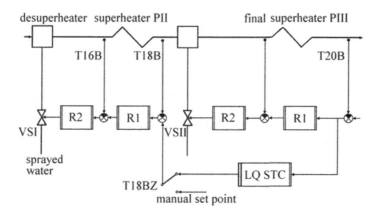

Figure 8.30. Diagram of adaptive control of steam temperature of the final two superheaters

$$e_p(k)^2 < K_1 R(k) \text{ or } e_p(k)^2 > K_2 R(k)$$

where $K_1 \sim 0.25$, $K_2 \sim 2.0$ and $R(k)$ denotes the estimate of the covariance matrix, then calculations of the estimates of parameters $\theta(k)$ or $\theta(k)$ and $R(k)$ were skipped.

8.4.3 Control Results

No identifications were made before the experiment and reasonable structural parameters had been chosen directly during tests – similarly to a previous application; the prediction error performance and parameter variances (*i.e.* uncertainty of parameters) and the quadratic criterion of control errors were compared.

(A) Control of steam temperature behind superheater PII.

The manipulated variable was in this case steam temperature behind the spray valve controlled by R2 – see Figure 8.29. The control experiments started with a SISO controller with model order $N = 3$. External measured variables were not used here – no sufficiently important permanent relations were found. (The influence of external variables on model prediction depends also on the sampling period, usually decreasing with increased period; a rather long period was used in this application.) The IST strategy with cautious variant and penalization of input differences was used for most experiments. The horizon $M = 3$ was used even with the IST strategy to respect possible changes in parameters. Exponential forgetting with conditional interruption of identification was used. Sampling period T_0 was set to eliminate the dynamics of the small analogue control loop and was taken in the range of 25 s to 50 s, without significant differences in control performance. The penalization of input differences was suitable for changing operation conditions. The LQ STC calculated the set point for the analogue slave control loop controlling the steam temperature behind the spray valve. Both types of model input $u(k)$ definition were tested, *i.e.* desired or measured value of manipulated variables were defined like $u(k)$ (see Note: Input definition in Section 8.3, case (A) and (B)). The control results were comparable. As the use of the measured manipulated variable seemed better from the point of view of prediction error and also avoided potential wind up in the case when the actuator's position is limited, the measured manipulated variable was mainly used as the input $u(k)$. Tests with an incremental model led to prediction and control error increases and therefore a such model was not applied. Typical control responses with and without LQ STC are documented as plant plotter records and shown in Figure 8.31 and Figure 8.32.

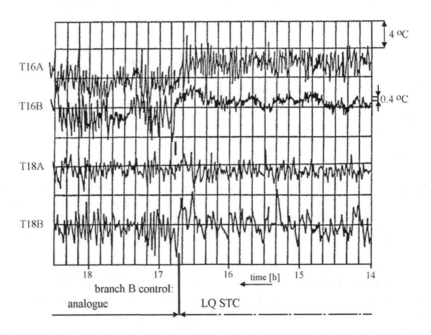

Figure 8.31. Steam temperature control of superheater PII. Branch B (temperature T18B) is controlled with analog control or LQ STC: model order $N = 3$, number of steps $M = 3$, $T_0 = 50$ s

Controlled steam temperatures behind the superheater T18B, and steam temperature behind the spray T16B, (*i.e.* manipulated variable for main con-

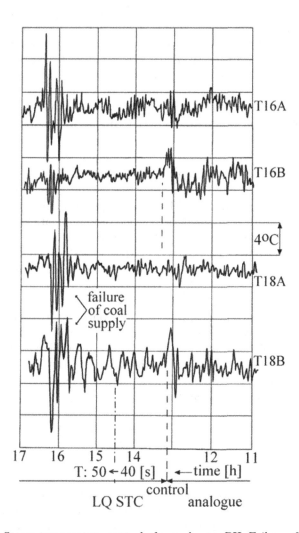

Figure 8.32. Steam temperature control of superheater PII. Failure of coal supply. Branch B (temperature T18B) is controlled using analogue control or LQ STC. LQ STC : model order $N = 3$, number of steps $M = 3$, change of sampling period T_0 from 40 to 50 s

troller) are shown, together with temperatures of branch "A" controlled only with standard analogue PID controllers. In Figure 8.32 the sampling period of LQ STC was switched from 40 to 50 s ; then a failure of the coal supply (of coal mill) caused oscillations, however, control errors were in a similar range for the usually better controlled branch A and for branch B controlled by LQ STC – in spite of the very long sampling period $T_0 = 50$ s of LQ STC.

A comparison of analogue and LQ STC control gives the following results: The application of LQ STC led to a significant decrease of changes in the manipulated variable (steam temperature behind the spray valve) with lower overshoots in the controlled variable and a decrease of the controlled and manipulated variable frequencies, resulting in an important heat stress decrease. The frequency of changes of the manipulated variable and their magnitude were decreased by approximately three times. At the same time, the range of control error remained approximately the same as for analogue control.

(B) Control of steam temperature of final two superheaters.

The manipulated variable was, in this case, steam temperature behind PII. The same structure of SISO LQ STC was used with sampling period 60 s. The input $u(k)$ was defined as the desired value of manipulated variable, *i.e.* the set point of the steam temperature behind PII. In this case the model used included the dynamics of the system PIII and PII with all inner analogue control loops. Application of the adaptive controller (Figure 8.31) resulted in a decrease of the overshoots in the temperature behind the PIII during the usual operating failures and in smoothing the spray valve control. From the comparison between steam temperature control on PIII of both branches (see T20B and T20A in Figure 8.33) it can be seen that with analogue control, overshoots of the right branch "B" are substantially larger than those on branch "A" in the case of sudden failures. The LQ STC on branch "B" significantly reduced the control deviation and overshoots, even in comparison with the usually better branch "A".

Figure 8.33 (from plant plotter records) shows responses of controlled steam temperatures behind the superheater PIII (T20A, T20B), temperature behind the evaporator TVYP, steam flow M4 and electric power NEL.

Conclusions

LQ STC improved control compared with the existing controllers. The second application in which another controller (here the LQ STC) was connected into the control loops of two superheaters was unusual and not previously tested. One adaptive controller plus 2*3 =6 analogue control loops with their dynamics and nonlinearities were involved. LQ STC was applied without any previous identification. Only about 3 hours were required to choose a proper LQ STC model structure and Q_u penalisation for both applications. The bursting effect (covariance wind up) does not take place, mainly due to excitation of the system by disturbances and also due to above mentioned conditional interruption of identification. About 20 to 30 % of data were not taken into the identification after the test on the small prediction error (for $K_1 = 0.25$). Satisfactory and efficient control was achieved with LQ STC also in the case of fast process changes and failures – in spite of longer sampling times (up to 50 or 60 s). Such long sampling times, impossible with digital PID controllers,

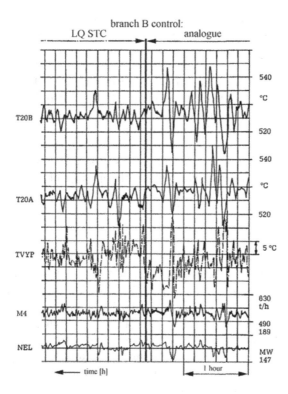

Figure 8.33. Steam temperature adaptive control of the final two superheaters PII and PIII. Branch B (temperature T20B) is controlled using analogue control or LQ STC: model order $N = 4$, number of steps $M = 3$, $T_0 = 60\,$s

were surprising to plant control engineers. This application pointed out also potential problems caused by the danger of wind-up effect in cascade control loops. Possible solutions are described in [64] and [65].

8.5 Steam Pressure Control of Drum Boiler in Power Plant

This application was similar to the application described in Section 8.3 but here analogue controllers ran, however, not fully successfully under large changes and process failures, this bring the reason why LQ STC was applied.

8.5.1 The Technology and Controlled Loops

The LQ STC was applied for steam pressure control with additional adaptive control of the combustion air on a coal-fuel-fired drum boiler 220 t/h, delivering steam with three other similar boilers into a common collector in a power plant. On the boiler there are 11 coal feeders; the number of coal feeders used varies from 5 up to 11, according to load. The combustion air intake is controlled by turning the control rims of two fans. The powdered coal intake is measured only indirectly by the positions of actuators on the speed changing units and depends also on the state of the coal bunkers and the quality of milled coal, the calorific value of which varies. Therefore the dependence of coal intake on feeder speed is time variable. The quality of the combustion process depends not only on the number of feeders used but also on their choice.

8.5.2 Programming Aspects

LQ STC was programmed as a MIMO controller with measured external disturbances similarly to both previous applications. Penalization of input or input increments could be used. Not only a constant value of penalization Q_u but also a variable value was applied. This value changed if the input (*i.e.* manipulated variable) reached its limits [119] and also according to the input variance (to automate tuning under conditions of output variance) – see [135]. Additional adaptive control of the combustion air, which was always used with LQ STC of steam pressure, was based on the estimated parameters of the process model created by this LQ STC.

8.5.3 Control Results

A SISO controller with two measurable disturbances and with IST strategy and penalization of input increments was used. The multi-step criterion with $M = 3$ was used as standard. However to demonstrate the difference between multi-step and single-step criterion, the following experiment was made. See Figure 8.34, where the influence of different values Q_u and applied control strategies are recorded; in Figure 8.34 variable y denotes controlled steam pressure, u manipulated variable (speed of coal feeders limited in the range $umin, umax$) and v steam flow used as the measured disturbance. The multi-step IST strategy was significantly better; the simple single-step criterion was unsatisfactory (manipulated variable was often limited and remained on its stops, controlled variable oscillated wildly, system was not correctly excited and parameter estimates and prediction error performance were wrong).

The variable input penalization used here improved control behaviour, avoided wind-up when the actuator was limited and removed tuning of Q_u. To eliminate the wind-up effect the corrector described in Section 4.5.2, Equation (4.86) and in [64, 65], was applied – together with variable penalization

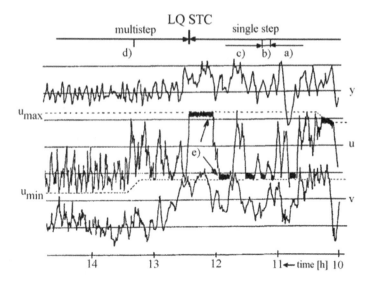

Figure 8.34. Steam pressure control of drum boiler 220 t/h. Comparison of single step v. multistep cost function. Change of penalization: a/ $Q_u = 0.0008$, b/ $Q_u = 0.008$, c/ $Q_u = 0.004$, d/ $T_0 : 40\,\text{s} \rightarrow 30\text{s}$, y denotes controlled steam pressure, u speed of coal feeders (manipulated variable), and v steam flow used as measured disturbance

increase if the actuator limited – see [119]. Again, the application of LQ STC improved control especially under large and fast changes and process failures (of coal mill *etc.*), which the use of PID controllers was not able to cope with.

References

1. R. Bellman, *Adaptive Control - A Guided Tour*. Princeton University Press, 1961.
2. Y. Z. Tsypkin, *Adaptation and Learning in Automatic Systems*. New York: Academic Press, 1971.
3. K. J. Åström, "Theory and applications of adaptive control," *Automatica*, vol. 19, no. 5, pp. 471–486, 1983.
4. G. C. Goodwin and K. S. Sinn, *Adaptive Filtering, Prediction and Control*. Englewood Cliffs, New Jersey: Prentice Hall, 1984.
5. K. J. Åström and B. Wittenmark, *Adaptive Control*. Reading, Massachusetts: Addison-Wesley Publishing Company, 1989.
6. P. Wellstead and M. Zarrop, *Self-tuning Systems*. Chichester: John Wiley & Sons, 1991.
7. N. M. Filatov and H. Unbehauen, *Adaptive Dual Control: Theory and Applications*. Berlin Heidelberg New York: Springer-Verlag, 2004.
8. J. Maršík and V. Strejc, "Application of identification-free algorithms for adaptive control," *Automatica*, vol. 25, pp. 225–228, 1989.
9. K. J. Åström and T. Hägglund, *Automatic Tuning of PID Controllers*. Research Triangle Park, North Carolina: Instrument Society of America, 1988.
10. K. J. Åström and T. Hägglund, "Automatic tuning of simple regulators with specifications on phase and amplitude margins," *Automatica*, vol. 20, pp. 645–651, 1984.
11. K. J. Åström and T. Hägglund, *PID Controllers: Theory, Design and Tuning*. Research Triangle Park, North Carolina: Instrument Society of America, 1995. 2nd Edition.
12. T. W. Kraus and T. J. Myron, "Self-tuning PID controller uses pattern recognition approach," *Control Engineering*, vol. June, pp. 106–111, 1984.
13. Y. Nishikawa, N. Sannomia, T. Ohta, and H. Tanaka, "A method for auto-tuning of PID control parameters," *Automatica*, vol. 20, pp. 321–332, 1984.
14. K. J. Åström, C. C. Hang, P. Persson, and W. K. Ho, "Towards intelligent PID control," *Automatica*, vol. 28, pp. 1–9, 1992.
15. W. K. Ho, C. C. Hang, and L. S. Cao, "Tuning of PID contollers based on gain and phase margin specifications," *Automatica*, vol. 31, pp. 497–502, 1995.
16. I. D. Landau, *Adaptive Control – the Model Reference Approach*. New York: Marcel Dekker, 1979.

17. P. A. Ioannou, *Robust Adaptive Control*. Englewood Cliffs, New Jersey: Prentice Hall, 1996.

18. A. Feldbaum, "Theory of dual control," *Autom. Remote Control*, vol. 21, no. 9, 1960.

19. R. E. Kalman, "Design of a self optimizing control system," *Trans. ASME*, vol. 80, pp. 481–492, 1958.

20. V. Peterka, "Adaptive digital regulation of noisy systems," in *Preprints of the 2nd IFAC Symposium on Identification and Process Parameter Estimation*, p. 6.2, Prague: ÚTIA ČSAV, 1970.

21. K. J. Åström and B. Wittenmark, "On self-tuning regulators," *Automatica*, vol. 9, pp. 185–199, 1973.

22. D. W. Clarke and P. J. Gawthrop, "Self-tuning controller," *Proc. IEE*, vol. 122, pp. 929–934, 1975.

23. D. W. Clarke and P. J. Gawthrop, "Self-tuning control," *Proc. IEE*, vol. 126, pp. 633–640, 1979.

24. V. Peterka, "Predictor-based self-tuning control," *Automatica*, vol. 20, no. 1, pp. 39–50, 1984. Reprinted in : *Adaptive Methods for Control System Design, Editor M.M. Gupta, IEEE Press, 1986*.

25. M. Kárný, A. Halousková, J. Böhm, R. Kulhavý, and P. Nedoma, "Design of linear quadratic adaptive control: Theory and algorithms for practice," *Kybernetika*, vol. 21, 1985. Supplement to No. 3, 4 ,5, 6.

26. J. Böhm, A. Halousková, M. Kárný, and V. Peterka, "Simple LQ self-tuning controllers," in *Preprints of 9th IFAC World Congress*, vol. 8, pp. 171–176, 1984.

27. P. E. Wellstead, J. M. Edmunds, D. I. Prager, and P. M. Zanker, "Pole zero assigment self-tuning regulator," *International Journal of Control*, vol. 30, pp. 1–26, 1979.

28. K. J. Åström and B. Wittenmark, "Self-tuning controller based on pole-zero placement," *IEE-Procedings D*, vol. 127, pp. 120–130, 1980.

29. D. W. Clarke, C. Mohtadi, and P. S. Tuffs, "Generalized predictive control – part i. the basic algorithm," *Automatica*, vol. 23, pp. 137–148, 1987.

30. D. Clarke, C. Mohtadi, and P. Tuffs, "Generalized predictive control," *Automatica*, vol. 23, no. 2, pp. 137–160, 1987.

31. P. J. Gawthrop, "Hybrid self-tuning control," *IEE-Procedings D*, vol. 127, pp. 229–236, 1980.

32. C. Bányász and L. Keviczky, "Direct methods for self-tuning PID regulators," in *Proc. 6th IFAC Symposium on Identification and System Parameter Estimation*, pp. 1249–1254, 1982.

33. V. Bobál, J. Böhm, and R. Prokop, "Practical aspects of self-tuning controllers," *International Journal of Adaptive Control and Signal Processing*, vol. 13, pp. 671–690, 1999.

34. R. Kulhavý and F. J. Kraus, "On duality of regularized exponential and linear forgetting," *Automatica*, vol. 32, pp. 1403–1415, 1996.

35. J.Lampinen and I. Zelinka, *New Ideas in Optimization - Mechanical Engineering Design Optimization by Differential Evolution*, vol. 1. McGraw-Hill, 1999.

36. I. Zelinka, "Evolutionary identification of predictive models," in *ISF' 2000, The 20th International Symposium on Forecasting*, Lisbon, Portugal: International Institute of Forecasters, 2000.

37. C. Kar, *Practical Applications of Computational Intelligence for Adaptive Control*. CRC Press, 1999.

38. R. Ballini and F. Zuben, *Application of Neural Networks to Adaptive Control of Nonlinear Systems.* Research Studies Press Ltd, 1997.
39. H. Hellendoorn and D. Driankov, *Fuzzy Model Identification.* Springer Verlag, 1997.
40. J. Böhm and M. Kárný, "Merging of user's knowledge into self-tuning controllers," in *Preprints of 4th IFAC Symposium Adaptive Control and Signal Processing ACASP'92* (I. Landau, L. Dugard, and M. M'Saad, eds), vol. 2, pp. 427–432, Grenoble: Academic Press, 1992.
41. R. Kulhavý and M. B. Zarrop, "On general concept of forgetting," *International Journal of Control,* vol. 58, no. 4, pp. 905–924, 1993.
42. L. Ljung, *System Identification: Theory for the User.* London: Prentice-Hall, 1987.
43. T. Söderström and P. Stoica, *System Identification.* Englewood Cliffs, New Jersey: Prentice-Hall, 1989.
44. L. Ljung and T. Söderström, *Theory and Practice of Recursive Identification.* Cambridge, Massachusetts: MIT Press, 1983.
45. V. Strejc, "Least squares parameter estimation," *Automatica,* vol. 16, pp. 535–550, 1980.
46. V. Peterka, "A square-root filter for real-time multivariable regression," *Kybernetika,* vol. 11, pp. 53–67, 1975.
47. G. Bierman, *Factorization Methods for Discrete Sequential Estimation.* New York: Academic Press, 1977.
48. V. Peterka, "Real-time parameter estimation and output prediction for ARMA-type system models," *Kybernetika,* vol. 17, pp. 526–533, 1981.
49. R. Kulhavý, "Directional tracking of regression-type model parameters," in *Preprints of the 2nd IFAC Workshop on Adaptive Systems in Control and Signal Processing* (Lund, Sweden), pp. 97–102, 1986.
50. R. Kulhavý, "Restricted exponential forgetting in real-time identification," *Automatica,* vol. 23, no. 5, pp. 589–600, 1987.
51. M. Kubalčík and V. Bobál, "Adaptive control of coupled drives apparatus based on polynomial approach," in *Proceedings of the IEEE International Conference on Control Applications 2002,* pp. 594–599, 2002.
52. V. Bobál, P. Navrátil, P. Dostál, and M. Sysel, "Delta self-tuning control MIMO systems: comparison 1dof and 2dof configurations," in *Proc. of IASTED International Conference Circuits, Signals and Systems,* pp. 5–10, 2003.
53. P. Dostál, V. Bobál, and M. Blaha, "One approach to adaptive control of nonlinear processes," in *Preprints of IFAC Workshop on Adaptation and Learning in Control and Signal Processing ALCOSP* (Cernobbio-Como, Italy), pp. 407–412, 2001.
54. R. Isermann, *Digital Control Systems.* Berlin, Heidelberg, New York: Springer-Verlag, 1991. 2nd Edition.
55. A. Niederliński, *Digital Systems for Control of Technological Processes: Hardware and Software.* Praha: SNTL, 1984 (in Czech).
56. Y. Takahashi, C. Chan, and D. Auslander, "Parametereinstellung bei linearen DDC-algorithmen," *Regelunstechnik und Prozessdatenverarbeitung,* vol. 19, pp. 237–284, 1971.
57. C. C. Hang, K. J. Åström, and W. K. Ho, "Refinements of the Ziegler–Nichols tuning formula," *IEE Procedings-D,* vol. 138, pp. 111–118, 1991.
58. K. J. Åström, C. C. Hang, P. Persson, and W. K. Ho, "Towards intelligent PID control," *Automatica,* vol. 28, pp. 1–9, 1992.

59. V. Bobál, J. Macháček, and R. Prokop, "Tuning of digital PID controllers based on Ziegler-Nichols method," in *Proc. of the 2nd IFAC Workshop on New Trends in Design of Control Systems* (Smolenice, Slovakia), pp. 133–138, 1997.

60. J. Šindelář, "Adaptation of control system structure to the course of controlling error," in *Proc. of the 2nd IFAC Symposium on System Sensitivity and Adaptivity* (Dubrovník), 1978.

61. J. Šindelář, *Control Systems with Discontinuously Variable Structure*. Praha: Academia, 1973 (in Czech).

62. P. D. Roberts, "Simple tuning of discrete PI and PID controllers," *Measurement and Control*, vol. 9, pp. 227–234, 1975.

63. F. B. Shinskey, *Process Control Systems*. New York: Mc Graw Hill, 1979.

64. J. Fessl and J. Jarkovský, "Cascade control using adaptive controllers with on-line identification: Some problems and solutions," in *Proc. of the 10th IFAC World Congress, Vol. 10* (Munich), pp. 59–64, 1987.

65. J. Fessl and J. Jarkovský, "Cascade control of superheated steam temperature with lq self-tuning controllerss," in *Proc. of the 8th IFAC Symposium on Identification and System Parameter Estimation* (Beijing), 1988.

66. A. H. Glattfelder and W. Schaufelberger, "Stability analysis of single loop control systems with saturation and anti-reset wind-up circuits," *IEEE Trans. Automatic Control*, vol. 28, pp. 1074–1081, 1983.

67. L. Rundquist, *Anti-reset Windup for PID Controllers. Report LUTFD2/(TRFT-1033)/ 1-143*. Lund, Sweden: Lund Inst. of Tech., 1991.

68. L. Rundquist, "Anti-reset windup for PID controllers," in *Proc. of the 11th IFAC World Congress, Vol. 8*, (Tallinn), pp. 146–151, 1991.

69. K. J. Åström and B. Wittenmark, *Computer-Controlled Systems: Theory and Design*. New Jersey: Englewood Cliffs Prentice Hall, 1984.

70. D. W. Clarke, "PID algorithms and their computer implementation," *Transactions of the Institute of Measurement and Control*, vol. 6, pp. 305–316, 1984.

71. R. M. Hanus, M. Kinneart, and J. Henrotte, "Conditioning technique on general anti-windup and bumpless transfer method," *Automatica*, vol. 23, pp. 729–739, 1987.

72. B. Šulc, "Integral wind-up in control and system simulation," *Control Engineering Solutions. A Practical Approach. IEE Control Engineering Series*, vol. 54, pp. 61–76, 1997.

73. B. H. Bristol, "Designing and programming control algorithms for DDC systems," *Control Engineering*, vol. 1, pp. 24–26, 1977.

74. A. B. Corripio and P. M. Tomkins, "Industrial application of self-tuning feedback control algorithhm," *ISA Transactions*, vol. 20, pp. 3–10, 1981.

75. D. B. Dahlin, "Designing and tuning digital controllers," *Inst. Control Systems*, vol. 42, pp. 77–83, 1968.

76. K. C. Chiu, A. B. Corripio, and C. L. Smith, "Digital control algorithms, part 1, Dahlin algorithm," *Inst. Control Systems*, vol. 47, pp. 57–59, 1973.

77. B. Wittenmark, *Self-tuning PID-controllers based on pole placement. Report LUTFD2/(TRFT-7179)/ 1-037*. Lund, Sweden: Lund Inst. of Technology, 1979.

78. R. Ortega and R. Kelly, "PID self-tuners: Some theoretical and practical aspects," *IEEE Trans. Ind. Electron.*, vol. 31, pp. 332–338, 1984.

79. J. H. Kim and K. K. Choi, "Self-tuning discrete PID controller," *IEEE Trans. Automatic Control*, vol. 34, pp. 298–300, 1987.

80. J. H. Kim and K. K. Choi, "Design of direct pole placement PID self-tuners," *IEEE Trans. Automatic Control*, vol. 34, pp. 351–356, 1987.

81. R. Kofahl and R. Isermann, "A simple method for automatic tuning of PID-controllers based on process parameter estimation," in *Proc. American Control Conference* (Boston), pp. 1143–1148, 1985.

82. F. Radke and R. Isermann, "A parameter-adaptive PID controller with step-wise parameter optimization," *Automatica*, vol. 23, pp. 449–457, 1987.

83. S. Tzaseftas and G. Kapsiotis, "PID self-tuning control combining pole-placement and parameter optimization features," in *Proc. IMACS/IFAC Second International Symposium on Mathematical and Intelligent Models in System Simulation, Vol. II* (Brussels), pp. 30–40, 1993.

84. P. J. Gawthrop, *Using the self-tuning controller to tune PID regulators. Report No. CE/T/2.* Brighton: School of Appl. Sci., Univ. Sussex, 1982.

85. P. Neuman, "Industrial adaptive regulator with discontinuously structure," in *Proc. of IFAC Workshop Evaluation of Adaptive Control Strategies in Industrial Applications* (Tbilisi), pp. 281–288, 1989.

86. M. Alexík, "Adaptive self-tuning algorithm based on continuous synthesis," in *Proc. of the 2nd IFAC Workshop on New Trends in Design of Control Systems*, (Smolenice, Slovakia), pp. 481–486, 1997.

87. C. Bányász and L. Keviczky, "Direct methods for self-tuning PID controllers," in *Proc. of the 6th IFAC Symposium on Identification and System Parameter Estimation*, (Washington), pp. 1249–1254, 1982.

88. C. Bányász and L. Keviczky, "Design of adaptive PID regulators based on recursive estimation of the process parameters," *Journal of Process Control*, vol. 3, pp. 53–59, 1993.

89. J. Böhm, A. Halousková, M. Kárný, and V. Peterka, "Simple lq self-tuning controllers," in *Proc. of the 9th Triennial World Congress of IFAC* (Budapest), pp. 961–966, 1984.

90. M. Vítečková, A. Víteček, and L. Smutný, "Simple PI and PID controllers tuning for monotone self-regulating plants," in *Proc. of the IFAC Workshop Digital Control: Past, Present and Future of PID Control* (Terrassa, Spain), pp. 259–264, 2000.

91. M. Vítečková, A. Víteček, and L. Smutný, "Controller tuning for controlled plants with time delay," in *Proc. of the IFAC Workshop Digital Control: Past, Present and Future of PID Control*, (Terrassa, Spain), pp. 253–258, 2000.

92. V. Bobál, P. Dostál, J. Macháček, and M. Vítečková, "Self-tuning PID controllers based on dynamics inversion method," in *Proc. of the IFAC Workshop Digital Control: Past, Present and Future of PID Control*, (Terrassa, Spain), pp. 167–172, 2000.

93. V. Bobál, *Auto-Tuning of Digital PID Controllers Using Recursive Identification. Report ESR 9409.* Bochum: Faculty of Electrical Engineering, Ruhr-University, 1994.

94. V. Bobál, "Self-tuning Ziegler-Nichols PID controller," *International Journal of Adaptive Control and Signal Processing*, vol. 9, pp. 213–226, 1995.

95. V. Bobál, "Robust self-tuning PID controller," in *Proc: 1st IFAC Workshop on New Trends in Design of Control Systems* (Smolenice, Slovakia), pp. 312–317, 1994.

96. V. Bobál, J. Macháček, and R. Prokop, "Practical tuning of industrial PID controllers," in *Proc. of the IFAC-IFIP-IMACS Conference Control of Industrial Systems* (University of Belfort), pp. 37–42, 1997.

97. J. Macháček and V. Bobál, "Adaptive PID controller with time delay," in *Proc. of the European Control Conference ECC99* (Karlsruhe), pp. paper BA – 12 – 5, 1999.

98. J. Macháček and V. Bobál, "Adaptive PID controller with on-line identification," *Journal of Electrical Engineering*, vol. 53, pp. 233–240, 2002.

99. C. Bányász, J. Hetthessy, and L. Keviczky, "An adaptive PID regulator dedicated for microprocessor based compact controllers," in *Proc. of the 7th IFAC Symposium on Identification and System Parameter Estimation*, (York), pp. 1299–1304, 1985.

100. L. Keviczky and C. Bányász, "A completely adaptive PID regulator," in *Proc. of the 8th IFAC Symposium on Identification and System Parameter Estimation* (Beijing), pp. 91–97, 1988.

101. K. J. Åström and B. Wittenmark, "Simple self-tuning controllers," in *Proc. Symposium on Methods and Applications in Adaptive Control* (Bochum), pp. 21–30, 1980.

102. V. Bobál, M. Kubalčík, and M. Úlehla, "Auto-tuning of digital PID controllers using recursive identification," in *Proc. of the 5th IFAC Symposium on Adaptive Systems in Control and Signal Processing* (Budapest), pp. 384–389, 1995.

103. J. G. Ziegler and N. B. Nichols, "Optimum settings for automatic controllers," *Trans. ASME*, vol. 64, pp. 759–768, 1942.

104. H. Cui and E. W. Jacobsen, "Performance limitations in decentralized control," *Journal of Process Control*, vol. 12, pp. 485–494, 2002.

105. V. Kučera, *Discrete Linear Control: The Polynomial Equation Approach.* Chichester: John Wiley, 1979.

106. V. Kučera, *Analysis and Design of Discrete Linear Control Systems.* London: Prentice Hall, 1991.

107. M. Vidyasagar, *Control System Synthesis: A Factorization Approach.* Cambridge MA, MIT Press, 1985.

108. V. Kučera, "Equations in control a survey," *Automatica*, vol. 29, pp. 1361–1375, 1993.

109. M. Šebek, *The Polynomial Toolbox 2 Manual.* Prague: Polyx, 1999.

110. V. Kučera, "A dead-beat servo problem," *Int. J. Control*, vol. 32, pp. 107–113, 1980.

111. M. Šebek, "Multivariable dead-beat servo problem," *Kybernetika*, vol. 16, pp. 442–453, 1980.

112. S. M. Shinners, *Advanced Modern Control System Theory and Design.* New York: John Wiley, 1998.

113. M. Šebek and V. Kučera, "Polynomial approach to quadratic tracking in discrete linear systems," *IEEE Trans. Automatic Control*, vol. AC-27, pp. 1248–1250, 1982.

114. J. Böhm, M. Kárný, and A. Halousková, "LQ optimization with irregular input-output sampling: Algorithmic and complexity aspects," in *Preprints of the European IEEE Workshop CMP'94* (L. Kulhavá, M. Kárný, and K. Warwick, eds), pp. 265–268, Prague: ÚTIA AVČR, 1994.

115. J. Böhm, "On algorithmic problems related to multivariate adaptive LQG controllers," in *Symposium on Control, Optimization and Supervision, CESA'96*, pp. 310–314, 1996.

116. I. D. Landau, D. Rey, A. Karimi, A. Voda, and A. Franco, "A flexible trans-mition system as a benchmark for robust digital control," *European Journal of Control*, vol. 1, pp. 77–96, 1995.

117. B. Wittenmark, "An active suboptimal dual controller for systems with stochastic parameters," *Automatic Control Theory and Application*, vol. 3, no. 1, pp. 13–19, 1975.

118. V. Peterka, "Digital control of processes with random disturbances and uncertain characteristics," Technical Report, ÚTIA AVČR, POB 18, 18208 Prague 8, CR, 1975 in Czech.

119. J. Böhm, "LQ self-tuners with signal level constraints," in *Preprints of the 7th IFAC/IFIP Symposium on Identification and System Parameter Estimation*, vol. 1, pp. 131–137, 1985.

120. J. Böhm and M. Kárný, "Self-tuning regulators with restricted inputs," *Kybernetika*, vol. 18, no. 6, pp. 529–544, 1982.

121. R. Bitmead, M. Gevers, and V. Wertz, *Adaptive Optimal Control. The Thinking Man's GPC*. Prentice Hall, 1990.

122. C. E. de Souza, "Monotonicity and stabilizability results for the solutions of the Riccati difference equation," in *Preprints of the Workshop on the Riccati Equation in Control, Systems and Signals*, pp. 38–41, 1989.

123. W. H. Kwon and A. E. Pearson, "On feedback stabilization of time-varying discrete linear systems," *IEEETransactions on Automatic Control*, vol. AC-23, pp. 479–481, 1978.

124. J. Böhm, "The set point control and offset compensation in the discrete LQ adaptive control," *Problems of Control and Information Theory*, vol. 17, no. 3, pp. 33–46, 1988.

125. B. D. O. Anderson and J. Moore, *Optimal Control. Linear Quadratic Method*. London: Prentice-Hall, 1989.

126. P. Nedoma, and M. Kárný and J. Böhm, "Designer : Preliminary tuning of adaptive controllers," in *Proceeding of the 2nd scientific–technical Conference PROCESS CONTROL, Horní Bečva, 1996* (J. Krejčí, ed.), pp. 225–228, 1996.

127. P. Nedoma, and M. Kárný and J. Böhm, *ABET: Adaptive Bayesian Estimation Toolbox for MATLAB*. Prague, Czech Republic: ÚTIA AV ČR, 1996.

128. J. Böhm, "Experiments with LQ adaptive controller in the heat exchanger station at STU Bratislava," Technical Report 1919, ÚTIA AVČR, P.O.Box 18, 182 08 Prague, Czech Republic, 1998.

129. J. Fessl, "An application of multivariable self-tuning regulators to drum boiler control," *Automatika*, vol. 22, pp. 581–585, 1986.

130. V. Bobál, J. Böhm, and P. Chalupa, "Matlab-toolbox for CAD of simple self-tuning controllers," in *Proc. of the 7th IFAC Workshop on Adaptation and Learning in Control and Signal Processing* (Cernobbio-Como, Italy), pp. 273–278, 2001.

131. V. Bobál and P. Chalupa, *Self-tuning Controllers Simulink Library*. Zlín: Tomas Bata University in Zlín, Faculty of Technology, 2002. http://www.utb.cz/stctool/.

132. P. Chalupa, *Discrete Decentralized Control Systems. Ph.D. Thesis*. Zlín: Tomas Bata University in Zlín, Faculty of Technology, 2003.

133. P. Zelinka, B. Rohal'-Ilkiv, and A. Kuznetsov, "Experimental verification of stabilizing predictive control," *Control Engineering Practice*, vol. 7, pp. 601–610, 1999.

134. J. Fessl and J. Jarkovský, "Steam superheater control via self-tuning regula-
      tor," in *Preprints of IFAC Symposium on Digital Computer Applications to
      Process Control*, Vienna: Academic Press, 1985.
135. J. Fessl, "LQ self-tuning controllers with varying cost function weight," in
      *Preprints of Proc. of IFAC Symposium ITAC 91 on Intelligent Tuning and
      Adaptive Control*, Singapore: Academic Press, 1991.

Index

Printing and Binding: Strauss GmbH, Mörlenbach